The Ravens of Snover Canyon

The Ravens of Snover Canyon

● ● ●

Philip I. Moynihan

ISBN-13: 9781532945052
ISBN-10: 1532945051
Library of Congress Control Number: 2016907256
CreateSpace Independent Publishing Platform
North Charleston, South Carolina

To Freya and Tyr who were responsible for this story

Contents

CHAPTER 1

Enter the Raven

● ● ●

I GUESS IF ONE WERE ever to ask me if I had a favorite bird, I would have to say it was the raven. There is a certain imposing mystique about the raven, an aura of reverence that commands respect.

I haven't always had that impression. As I suspect is the case with most people, at first I knew very little about ravens and had given them very little thought overall. But a few years ago, I had the fortunate experience to have befriended a mated pair of wild ravens that had decided to make their home in Snover Canyon, an extremely rugged wilderness area located within the foothills of the mountains north of Los Angeles in Southern California, adjacent to where I live. The more I came to know these magnificent and complex creatures, the more in awe I became of them. I observed how rational they seemed when going about their daily lives: how they would find food, how they would play in the sky, and how they would sit together on a tree limb, where very often one would place its foot over its mate's, as if holding hands. I watched them raise their young ones and teach their little fledglings the skills they would need to survive in their new world. I shared their lives through their good times and their tragedies. I came to know them, as they came to know me. This tale is a detailed accounting of their story.

The raven has been a figure of mystery and legend in human cultures and has found its way into folklore and imaginations since the first hunter-gatherer band huddled around the fire pit telling stories.[1] This virtual symbiotic relationship between our two species has greatly enriched folktales, where the raven is either revered as a divine messenger or reviled as a mischievous destroyer. Not only has the raven been a prominent player in European mythology, but it was also seen as a messenger and oracle in ancient Asia and as a creator and a herald of divine origin among natives in the Americas. In the folklore of many European cultures, the raven portends ill will as a harbinger of death. In France, ravens were thought to be the souls of wicked priests. The ghosts of murdered people manifested themselves as ravens in Sweden, while in Germany they were associated with the souls of the damned. In other parts of the world, their numbers were so extensive that people saw them as pests and sought their eradication. Our two species have coexisted and interacted with each other for thousands of years.[2]

In Norse mythology, as the story goes, the realm of Asgard was ruled by the All-father, King Odin, and his wife, Frigg, while

---

1 John M. Marzluff and Tony Angell, *In the Company of Crows and Ravens* (New Haven: Yale University Press, 2005), 1–35.
2 J. Chappell, "Living with the Trickster: Crows, Ravens, and Human Culture," *PloS Biology* 4, no. 1 (2006): e14.

the gods went about their daily travails and conflicts, which mirrored the human condition of that time. But Odin, having been blinded in one eye when he exchanged it for wisdom in Mímir's well and also being prone to forgetfulness, depended upon his two ravens, who would fly daily throughout the world and return with information of events, which they would share with him. The ravens were named Huginn (which in the Old Norse language meant "thought") and Muninn ("memory"). Hence, virtually all renditions of Odin depict him with only one eye and sitting on his throne while overseeing the affairs of Asgard in the presence of the two ravens and with his two wolves, Geri and Freki, at his feet.

Even Greek mythology references the raven. The ancient Greeks associated ravens with Apollo and considered them to be his messengers. One story holds that Apollo was suspicious that his lover, Coronis, was being unfaithful, and he sent a white raven as a spy to determine if this were true. When the raven returned with verification that confirmed his suspicions, Apollo became so enraged that the heat of his fury scorched the white feathers of the raven, turning them black. Hence, all ravens today are black.

Virtually everyone is familiar with the ravens at the Tower of London. Legend has it that ravens occupy a special position in England as guardians of the kingdom.[3] As long as there are ravens at the Tower of London, England will be safe from any

3 "Ravens of the Tower of London," *Wikipedia*, last modified May 13, 2016, https://en.wikipedia.org/wiki/Ravens_of_the_Tower_of_London.

foreign invader. But were the ravens to be removed, then the kingdom would fall.

One story has it that their removal was ordered by Charles II in response to the royal astronomer, who complained of their despoiling the telescopes and interfering with his observations at the Royal Observatory. But as the effort was initiated to remove the ravens, Charles was told of the legend. And since England was just recovering from the English Civil War, Charles had second thoughts and bowed to the superstition. Instead of removing the ravens from the grounds of the tower, he relocated the Royal Observatory to Greenwich. The year was 1675. Today, the Greenwich observatory is the site of the Prime Meridian, the geographic locus marking the zero-degree longitudinal reference for the entire planet—all because of ravens.

The continuity of the generations of ravens occupying the tower grounds was interrupted by World War II. Unfortunately, the majority of the tower ravens did not survive the shock of the Blitz during the war, leaving only one mated pair. But before the tower could be reopened to the public after the war, one of the ravens disappeared and was followed shortly thereafter by the other. This incident made the news, and several of the newspapers published the story that the empire would fall with the disappearance of the ravens. And as fate would have it, the British Empire was broken up shortly after World War II, confirming the legend to the more superstitious.

To this day, a group of eight ravens occupies the grounds of the tower—six are required, and two are in reserve. They are meticulously cared for and fed by the ravenmaster, wearing his official Beefeater uniform, all to the delight of the tourists.

While similar stories exist throughout the various cultures of the world, perhaps to me the most fascinating are those told by the natives of North America.[4] In these stories, the raven plays a predominant role in the creation mythologies of the native peoples, while at the same time assuming the role of the trickster. In the mythologies of the Tlingit people of the Pacific Northwest, the raven is both the creator of the world (the one who brought light into darkness) and at the same time depicted as selfish, conniving, and mischievous. One Tlingit creation story relates that long before humans, when all things of the world like mountains, fire, and wind were created by the Great Spirit, he had separated them all out and stored them individually in cedar boxes.[5] He then gave a box to each of the animals. When they opened the boxes, all things that made up the world spilled out. Mountains, rivers, winds, and plants came into being. But there was no light. The Great Spirit had given the one box that contained all the light of the world to the seagull, and the seagull refused to open it. The other animals

4 Peter Goodchild, *Raven Tales* (Chicago: Chicago Review Press, 1991), 1–144.

5 Wlliam Reid and Robert Bringhurst, *The Raven Steals the Light: Native American Tales* (Boston: Shambhala, 1996), 1–128.

summoned the raven and asked him to persuade the seagull to open the box and release the light. The raven agreed and set about to undertake this task. But through all the wiles the raven could muster—from begging and cajoling to flattery and trickery—to get the seagull to open the box, the seagull refused. Frustrated, the raven became very angry and jabbed a thorn into the seagull's foot, causing the seagull to scream in pain and drop the box. As the box tumbled down, the sun, moon, and all the stars spilled out, bringing light to the world.

The raven is a major player in the culture of the Haida mythology, another indigenous people of North America.[6] While the Haidas often refer to the raven as a trickster, that characteristic is actually intended as an introspective condition of one's inner self, a reflection of human nature. Hence, the raven is the principal character in numerous Haida stories, as the Haida people use these stories to teach moral lessons. By presenting the raven as greedy and mischievous, the stories make a moral counterpoint to demonstrate desired behavior.

This same mischievous raven, this magical transformer, is likewise a powerful creative force, and the raven as a creator also holds a dominant role in Haida mythology. The raven is said to have created the Haida homeland, the present-day Queen Charlotte Islands off the northern coast of British Columbia. It

---

6 Robert Bringhurst, *A Story as Sharp as a Knife: The Classical Haida Mythtellers and Their World* (Vancouver: Douglas & McIntyre, 2000), 13–50.

was the raven in one story who released the first humans from a cockleshell. The story has it that the raven was bored and, while looking for amusement, found a cockleshell on the beach. He noticed that there were beings trapped inside the shell and pecked on the shell to release them. These beings were the first men to enter the world. But intimidated and reluctant to emerge, they had to be coaxed out of the shell by the raven. The story continues with different variations, but one of the more common threads has it that the raven soon became bored with these new creatures and was going to put them back into the shell. But then he had an inspiration. He would have more fun, he thought, if he could find a similar shell of trapped women, free them, and be entertained by the interaction of the two groups. So this he set out to do. He found female humans trapped inside a citron shell, pecked on it to free them, and then flew up to a perch to watch the interaction between the men and women.

One of the richer stories of the Haida people is how the raven helped bring the sun, moon, stars, water, and fire to the world. The essence of this story is that long ago, an eagle was the guardian of these elements, but he hated humans so much that he kept these items hidden, forcing people to live in darkness without fire and fresh water. But the eagle had a beautiful daughter the raven fell in love with. Because the raven at the time was pure white, the daughter was charmed by his plumage and invited him into her father's lodge. Once inside the lodge, the raven saw the sun, moon, stars, and fresh water hanging on the wall, and a fire in the hearth. The temptation was too great. When no one was looking, he stole everything from the wall, grabbed a lump of fire, and flew out of the lodge. Once outside, he hung the sun in the sky, creating so much light that he was able to fly far and wide—much farther than he had ever flown before. When the sun set, he hung the moon in the sky and scattered the stars all around. Now there was moonlight and starlight by which he could keep on flying. And he was still carrying the fresh water and the stolen fire.

When he reached a place that looked just right, he dropped all of the water he was carrying. As the water spread over the ground, it became the source of all the lakes and streams of the world. But while he continued to fly on with the lump of fire in his bill, the smoke blew over his white feathers, turning them black. Suddenly his bill was scorched, and he dropped the fire. The fire landed on the rocks and quickly hid within them. The raven's beautiful white feathers were permanently blackened by the smoke, never to be white again. He would now forever be a black bird. As for the fire, whenever one strikes two stones together, sparks will still escape.

Because of the important role the raven plays in the mythologies of the North American cultures, its image is often depicted in the tribal crests. It is common, for example, to see the raven in a prominent position on totem poles.

There are countless other stories from virtually every culture of the world depicting the curious bond between humans and ravens. Although we share the planet with thousands of other birds, there does seem to exist a special, unexplained attraction between us and this mysterious, magnificent black bird.

CHAPTER 2

Call Me a Birdbrain, Please!

● ● ●

WHY HAS THERE BEEN THIS human fascination with ravens over all these millennia? One of the primary reasons, I suspect, is due to the raven's very high intelligence. With intelligence comes complex behavior, and with complex behavior come actions that are not specifically dedicated to self-preservation or preservation of the species. One also sees activity that can only be interpreted as mischievous or playful—acts that may merely be a response to boredom or a need for entertainment.

But where would one get the idea that a raven was intelligent? After all, it's only a bird. How smart could a bird be? This is a fair question, so let's look at this issue in a bit more detail. The raven belongs to the family of corvids, or Corvidae, which includes crows, magpies, jays, and rooks, among others.[7] And ravens are the most widely distributed geographically of all the corvids, in that they are found everywhere across the Northern

---

7 N. Clayton and N. Emery, "Corvid Cognition," *Current Biology* 15, no. 3 (2005): R80–1.

Hemisphere. It might surprise one to learn that corvids are considered to be the most intelligent of all birds, Psittaciformes (parrots) included, and furthermore rank among the most intelligent of all animals.[8] A raven's display of intelligence is the stuff of legends. Its brain development is on par with that of the great apes![9]

Let's explore this point for a moment. When considering brain size and its potential meaning relative to intelligence, one must standardize its mass to the body mass of the species being studied. At first blush, one might consider simply ratioing brain mass to body mass and comparing the results. However, there is a major divergence, for example, between the ratio of a 1.3-kilogram human brain to a human body and that of a 14-gram raven brain to a raven body. While this approach is necessary for the determination of relative intelligence, it is not sufficient when the goal is to compare cognitive ability across disparate species.

Brain weight does not vary linearly with body weight when comparing species. Were this so, then this simplistic approach would reveal that both mice and humans have a body weight that is forty times the brain weight, seeming to imply that mice and men are of equal intelligence. Unfortunately, a simple ratio

8 Candace Savage, *Bird Brains: The Intelligence of Crows, Ravens, Magpies and Jays* (Toronto: Douglas & McIntyre, 1995), 1–144.

9 James Owen, "Crows as Clever as Great Apes, Study Says," *National Geographic News*, December 2004, news.nationalgeographic.com/news/2004/12/1209_041209_crows_apes.html.

of brain mass to body mass without consideration of other physical differences between the two species distorts the outcome.

But if one factors into this ratio how body size relates to a species' anatomy and physiology (in a manner similar to that first suggested by Otto Snell in 1892 and refined later by others, culminating with H. J. Jerison in the late 1960s[10]), then one can develop what is called an *encephalization quotient*, which is a nonlinear relationship between brain and body mass that factors out distortions caused by body-size limits. The encephalization quotient is expressed as the ratio of an animal's actual cephalization factor over its expected cephalization factor for that animal's given weight, where cephalization can be thought of as the trending in an organism's evolution to concentrate sensory and neural functions in the forebrain, and thus increase intelligence. The power of the encephalization quotient—although not perfect—is that it provides a more rigorous standardization for comparing relative intelligence across species. The higher the encephalization quotient, the higher the probability of that species being more intelligent. When one normalizes a raven's brain for its body size, one discovers a brain that is considerably larger than expected.

The following table contains a small sampling of comparative encephalization quotients. When one notes where the

---

10 Michael K. Brett-Surman, Thomas R. Holtz, "The Complete Dinosaur," *Indiana University Press, Bloomington. Ind.* (2012): 191-208.

raven ranks in intelligence when compared with more familiar species,[11] one can readily understand why corvids are referred to as "feathered apes."

Encephalization Quotient for Select Species

Species	Encephalization Quotient
Human	7.44
Chimpanzee	2.49
Raven	2.49
Rhesus Monkey	2.09
Dog	1.17
Horse	0.86
Mouse	0.5

Although the raven has a brain—specifically a forebrain—of the same relative size and capacity as that of the great apes, one must be reminded that the avian and mammalian brains are structurally quite different. When one thinks of this difference from an evolutionary perspective, it makes considerable sense. A bird like a raven is a flight animal, and the total body weight that must be lifted while flying is critical. Mammals,

---

11 N. J. Emery, "Cognitive Ornithology: The Evolution of Avian Intelligence," *Philosophical Transactions of the Royal Society of London, Biological Sciences* 361, no. 1465 (January 2006): 23–43.

meanwhile, are fundamentally ground animals and so carrying the extra weight is less of a concern.

This fundamental difference in brain structure has severely hindered the study of bird intelligence. Intelligence was long thought to be a function of a well-developed cerebral cortex, as it is with mammals. However, birds have essentially no corresponding brain structure, leading neurologists to conclude for a long time that birds were primarily driven by instinct. And moreover, neurologists until well into the last century further concluded that birds likely had very little natural ability to learn. This preconceived notion derived from an erroneous view of avian brain mechanics severely impeded any interest in developing studies of bird cognition. Hence, avian intelligence has only recently been a subject of investigation.

The countless hours invested over decades in observing birds in their natural habitat have been restricted almost entirely to studying their behavior and habits. Because of the misconceived perception of their being fundamentally instinctual, virtually no effort was directed toward studying their intelligence. Meanwhile, primatologists who also put in countless hours in their studies over similar time spans have extensively investigated the cognitive traits of their subjects.

Without getting too technical, we will merely accept that neurologists who study brain physiology have patterned their paradigm on the mammalian brain and have mapped out its functions, both autonomous and intellectual, accordingly. Because a bird's brain does not conform to this configuration, it had been grossly misunderstood until fairly recently.

This dramatic difference in brain structure between these two animal classes, the corvids and mammals, seems to have nonetheless converged toward a common function wherein they appear to have developed similar cognitive processes for the same outcome.[12] This phenomenon is an example of *convergent evolution,* driven by the demands of a changing environment that forces a species to adapt for self-preservation and preservation of its kind. In this case, the necessity for each animal class to solve complex problems, both social and environmental, became the evolutionary driver.[13]

There have been recent reevaluations of the definitions of the various regions of the avian brain that have contributed considerably toward explaining how birds perform mental functions in a manner similar to that of primates.[14] For example, even with the radically different brain structure, there is evidence of similar neural connectivity. As researchers have investigated avian brains more closely, they have learned that the avian cerebrum consists of a relatively large pallium (the layers of gray and white matter that cover the upper surface of the cerebrum) and that this large pallial area of a bird performs functions similar to those of the

---

12 N. J. Emery and N. S. Clayton, "The Mentality of Crows: Convergent Evolution of Intelligence in Corvids and Apes," *Science* 306 (2004): 1903–7.
13 J. Fritz and K. Kotrschal, "Social Learning in Common Ravens, *Corvus corax,*" *Animal Behavior* 57 (1999): 785–93.
14 A. Reiner et al., "Songbirds and the Revised Avian Brain Nomenclature," *Annals of the New York Academy of Sciences* 1016 (2004): 77–108.

mammalian cerebral cortex. Furthermore, a bird's pallium is made up of smaller and more densely packed neurons than are found in a mammal's brain. As a result, these observations suggest that the higher concentration of smaller neurons within this region enables more synapses within a given volume.[15] More neurons lead to more effective connection densities, and the necessity to minimize brain mass for flight optimization led to the development of shorter connection lengths among various brain regions for minimal wiring. Hence, a more complex brain function can be packaged into a smaller volume. The speculation is that this increase in number of neurons may lead to a highly efficient neural architecture that could function in the same way as the mammalian cerebral cortex.

What might have driven this evolutionary convergence of two very different brain structures toward very similar intellectual capability? To begin with, corvids and apes emerged as species during approximately the same epoch, around 5 million years ago. This was a period of dramatic environmental and climatic instability. The Isthmus of Panama was formed around 3.5 million years ago, cutting off the flow of warm equatorial currents into the Atlantic region. This upheaval initiated an Atlantic cooling cycle and precipitated the subsequent series of ice ages. Animal species emerging during this period were faced with a need to continuously adapt to changing climates and food

---

15 Gerhard Roth and Ursula Dicke, "Evolution of the Brain and Intelligence," *Trends in Cognitive Sciences* 9 (May 2005): 250–7.

availability. This adaptation required innovation. Diets had to be modified for survival, and the generalist foragers were the most successful. Feeding strategies required cleverness when looking for food. Those who learned these valuable lessons survived and passed on their genes, and corvids and apes became masters.

The fact that a raven's diet is omnivorous has been a major factor in its success as a species. Ravens will feed on virtually anything from carrion to fruits and berries, to insects and small animals. Their diet is extremely variable. This adaptation contributed toward developing a larger brain. These innovative foraging strategies would have also contributed to the social dynamic of the species, spurring more complex social behavior.[16] The ability to think, react, and adapt to these ever-changing environments would have contributed significantly to the species' evolving intelligence.

Another contribution toward enhanced brain development is a species' relatively long developmental period from infancy to the time when it leaves the parents. This stage provides a longer interval for the offspring to learn from the parents. The young of both corvids and apes readily enjoy this advantage.

All of these preconditions possessed by both corvids and apes correlate with the evolution of higher-order cognitive ability. I should point out that the family Psittaciformes, which

---

16 A. Bond et al., "Social Complexity and Transitive Inference in Corvids," *Animal Behaviour* 65, no. 3 (2003): 479–87.

includes parrots and cockatoos, also experienced a very similar development that led to a comparable result.

There are a myriad of examples demonstrating the intelligence of ravens, and we'll consider several later. But one of their traits that warrants further discussion at this point is a brief mention of their problem-solving capability.[17] Observers have recorded numerous instances of ravens solving complex problems. Although those efforts are usually associated with securing food, the motivation should not detract from the applied methodology. The raven typically proceeds to the desired result without demonstrating a trial-and-error process to learn the solutions. Virtually all other species generally require some form of trial-and-error approach until they stumble upon the correct solutions. The raven, on the other hand, seems to study the problem at hand, formulate the solution in its mind, and then proceed to execute it.

One winter not long ago in Scandinavia, an ice fisherman experienced a puzzling anomaly. In the morning he would go out on the ice covering a lake near where he lived and cut a hole that was nominally a foot in diameter. He then would cut a small branch from a nearby tree, tie his fishing line to the center of the branch, and drop the baited hook into the water. He would at that point leave and go about his business of the morning, expecting to return and pull up a fish. This proceeded as a

---

17 J. Fritz and K. Kotrschal, "An Experimental Investigation of Insight in Common Ravens (Corvus corax)." *The Auk* 112(4) (1995): 994–1003.

daily routine until one day he returned and found that the line had been pulled up and, along with a barren hook, lay strewn alongside the hole. This transgression went on for several days. The fisherman suspected that someone was stealing his catch, so he mounted a hidden camera nearby. Upon checking the video the next day, he made an astounding discovery.

It seems that a raven had been watching this daily activity from an adjacent tree. It had noticed that when the fisherman pulled up the line, there was usually a fish attached. So after the fisherman left his assemblage unattended, the raven would fly down to the hole in the ice. He would first pull the small branch onto the ice, and then he would begin to pull up the fishing line. He would pull up a section of line, step on it, pull up another section, step on it again, and continue this process until the fish that was caught on the hook appeared. The raven would then pull the fish onto the ice and eat it. Indeed, someone was stealing the fisherman's catch, but that someone wasn't human. It was a raven! And he caught it all on video!

Another example of how ravens rapidly solve complex problems is the experiment where a raven correctly and very quickly disassembled a complex configuration of a transparent box within a box to get at the morsel at the very center. The boxes were designed to be disassembled, but the disassembly required a certain order. The first step required that a string attached to one surface of the outer box be pulled to open that side. Then other sides of the outer box could be opened to expose the inner box. The inner box was opened similarly. The raven that participated in this experiment had not seen how these boxes went together, yet it disassembled them so quickly that one had to pay close attention to keep up. This same experiment was offered to two dogs who never did solve it.

One of the more famous examples of demonstrated raven intelligence is the case where the raven was to retrieve a piece of meat suspended at the end of a string. Here, as with the case with the fisherman, it had to assess the problem at hand and then pull up part of the string, step on that part, pull up another part, step on it, and continue the process until it had the meat. Once again, the raven accomplished this feat after nothing more than looking at the string, apparently applying rational thought, and then flying to the branch and performing the task. This task was further complicated for the raven when the experimenter hung two strings, one with the meat and one with a rock attached. The test was to see how many tries it would take the raven to make the correct selection. It did so on the first try, and it continued to do so on the first try. It completely ignored the rock.

There was one more experiment that had to be attempted. This time the experimenter crossed the two strings so that from the vantage point of the branch it would appear that the rock was under the point where the meat was attached, and the meat was under that of the rock. Now what would the raven do? It flew over to the branch, looked down, and with its beak grabbed the correct string. The experimenter repeated this numerous times, and the raven never made a mistake.[18]

There have been many anecdotal accounts of the cleverness corvids have displayed when faced with the daunting task of cracking the hard shell of a nut to get at the meat inside. A crow has been videoed dropping a nut onto a busy crosswalk and waiting for the traffic light to change for cars to pass through, hoping that one may run over it and break the shell. When the light changed again, it flew the point where the nut was dropped and gathered up the exposed meat.

I witnessed a very similar incident, but this one involved a raven and occurred at a local general-aviation airport. Adjacent to the airport's runway runs a small river, in which freshwater mussels can be found. This raven had found one. As we were performing a preflight preparation of our airplane, I noticed a raven nearby flying to a height of approximately twenty feet and dropping something. When it repeated this twice, I started to pay closer attention. The raven was dropping a small shellfish onto the asphalt in an attempt to crack the hard shell. It would drop the shell and then fly down to inspect it by pecking at it to see if any progress had been made. When it found it was still unable to get to the meat, it would pick it up, fly back up to the same height, and drop it again. I watched it perform this

---

18 B. Heinrich, *Mind of the Raven* (New York: Harper Collins, 1999), 312–24.

action twice more and then we had to leave—unfortunately not knowing whether the raven was ever successful. But successful or not, it knew what it had to do to crack the shell—and it was determined!

One of the more fascinating stories I happened upon was an accounting of two ravens defending their nest and young ones from marauding humans by throwing rocks at them.[19] Two ravenologists studying local raptor populations had observed a pair of ravens frequenting a vertical crack located fairly high up the face of a cliff. When the ravens departed, the investigators ascended the cliff and found a ravens' nest containing six nestlings. After a short time of note taking, they began their descent. As they were starting down, the parents returned and started attacking them by flying at the interlopers while calling out loudly. Then the birds retreated to the top of the cliff directly above the two men, who had resumed their descent. Suddenly a golf-ball-size rock flew past the face of one of the researchers. Thinking that one of the ravens had kicked it loose, both looked up, only to see a raven perched with a rock in its beak. The raven then flipped its head to toss that rock toward the marauders. The two men immediately took shelter under an overhang and watched while the raven threw six more rocks at them, striking one of the researchers in the leg. The largest

---

19 S. James, "The Apparent Use of Rocks by a Raven in Nest Defense," *Condor* 78, no. 3 (1976): 409.

rock was around three inches in diameter and an inch thick, and showed evidence that it had been partially buried prior to being thrown. This rock throwing by the raven was obviously an intentional, overt act of a parent protecting its young.

So the next time a raven alights onto your balcony railing— or even the lowly crow, for that matter—look into its eyes when it looks back at you. Somebody's home! And if anyone ever calls you a birdbrain, thank him or her for the compliment.

CHAPTER 3

Our New Neighbors

● ● ●

ONE SATURDAY AFTERNOON IN THE early 1990s, my wife, Penny, and I were alerted to a mild knocking or thumping on our sliding glass door leading to our upstairs balcony. It was a bit of a muffled knock. Curious as to what this was, I got up to investigate. As I approached the door, I abruptly stopped. I couldn't believe what I was witnessing. At the foot of the door stood a very large black bird with its wings outstretched; it was banging them against the glass. And behind it, perched on the rail, was another large black bird, occasionally calling to the first as if to egg it on. As I studied the dynamics of this odd scene for a moment, it became immediately obvious to me that these large black birds were ravens.

Although our house is located directly adjacent a national forest, until this event we had never seen ravens flying around, or even crows. The area was rich in virtually every other form of wildlife, but this was our first visit from ravens.

So what was going on? Why was this raven slapping the glass door? As I watched things unfold a bit longer, it dawned on me what was likely taking place. A mated pair of ravens was new to this region and wanting to establish a territory. Everything was apparently going well for them until they flew by our house. Because we are located at the edge of the national forest, our house has an abundance of windows. And all of the windows, including the sliding glass doors, are comprised of double-paned, tinted glass. On a bright, sunny day when the sun is just right, the windows are very reflective, almost like mirrors. When the ravens flew by, they must have seen another pair of ravens in their territory and come down to investigate! Now this scene began to make sense. The raven at the glass door was challenging this very aggressive-acting interloper to the rights to this territory. His mate was perched on the rail giving him encouragement. She was clearly yelling down at him.

I watched this scene unfold for quite some time. The male would press up against the glass, beat it with his outspread wings, and then step back. His opponent obviously did the same thing but didn't give ground. Instead, he just stood there defiantly. So our hero attacked once more while his mate cheered him on. I watched this scenario for a few more cycles until I felt that he had really had enough, at which point I went to open the door. They both immediately flew away upon seeing me.

When I looked at the glass he had been beating against, I noticed he had left a faint dust outline of his body and outstretched wings. With that, I went back into the house and thought nothing more about it. I assumed this would be an isolated event and we had seen the last of them. But that was not to be the case.

Later that week when I returned home from work, I again noticed the dust outline of an outstretched bird on the glass door as before, but in a different location. I thought it was odd, but the ravens were nowhere to be found, so I immediately dismissed it. When I returned from work the following day, I noticed that the outside screen door of this same sliding glass door was now opened by several inches. This too was strange, since even with heavy winds, that screen doesn't move. I repositioned the screen door but again didn't give this anomaly much thought.

But when I returned home the third day, something new was definitely afoot. This time, not only had the screen door been moved (and moved more than it had been the day before), but its vertical rubber dust seal had also nearly a foot of material stripped off. Pieces of the seal were scattered along the balcony. And again on the glass was an imprint of a bird with outstretched wings. Now I started to pay attention. Apparently these birds were serious! I had to find some way to draw their attention away from the glass door before they did more damage or hurt themselves.

If the threat to their establishing their territory was their reflections, then I thought maybe we could give them something with a much clearer reflection that would draw them away from the glass door. As luck would have it, we had the ideal solution. My brother-in-law was learning the trade of cutting and installing mirrors and other glasswork in new-house construction. As a practice piece, he had assembled four mirror strips, approximately four inches wide and a foot and a half long, into a vertical column, and then mounted the assembly onto a square pedestal. Each corner was bonded with a quarter-round wood strip approximately one half inch across. As artwork goes, it definitely was not the most elegant piece, but cutting these pieces had helped him gain the necessary experience in glasswork assembly. And he had given us this particular piece.

I took it out onto the balcony and placed it in the vicinity of the glass door—but far enough away to avoid any confusion with the door. Once outside, it was obvious that the reflection from an actual mirrored surface was so much clearer than that from the tinted glass of the door. I was confident that this would attract our raven's attention and cause him to redirect his challenges. And I was right.

When I returned home from work the following evening, I was met with a most amazing sight. Our little piece of artwork was in shambles! Our raven had torn it apart! The mirror pieces were strewn along the balcony and were covered with splintered fragments of the wood quarter rounds that had held the original pieces together. In addition, there was no new evidence of any disturbance of either the glass door or the screen. And from that day forward, the territorial supremacy had been established, and the phantom challenger in the glass door was never revisited. The experiment worked.

When this incident occurred, I had not yet retired (I retired in 2007), and I knew nothing about ravens. So, over the next few days, as I watched our two ravens soar over their newly established territory, I said to myself, *These are our new neighbors. I guess we ought to get to know them!*

CHAPTER 4

Freya and Tyr

● ● ●

OUR NEW NEIGHBORS HAD NOW moved in and established themselves as members of the community. They carved out a fairly large area as their territory, including the entirety of a virtually impassible wilderness called Snover Canyon. Due to the very steep hillsides and ravines, and overall extremely rugged terrain that comprised their new home, it is difficult to estimate the size of their territory. But as a guess, it likely covered an area of about one square mile, if one were to look at it from the air. This was going to be their new home, which they would jealously guard against interlopers—specifically other ravens. When a mated pair of ravens establishes a territory, it is theirs and theirs alone. No other ravens are welcomed. And when ravens mate, they mate for life. Such is the way of the things in Ravenworld.

So, as I had promised myself, I set about getting to know them. While watching them go about their daily routines, it became obvious early on in our relationship that I was observing two beings that were more than mere songbirds. They were demonstrating a curiosity. They were aware of me in a manner that reflected interest, as if they were also studying me. They were cautious of me but not really intimidated. The more I observed this characteristic, the more I knew that I was dealing with special beings and that I definitely had to know more about them and their species.

But they each needed a name. At first I was thinking of calling them by the same names as Odin's two ravens: Thought and Memory. But then it occurred to me that such names would probably be too abstract. They needed more down-to-earth names but still something not too common. Names from the pantheon of Norse mythology seemed most appropriate. So I decided to call the male Tyr after the Old Norse god of war and to call the female Freya after the Norse goddess of love and beauty. At this point in my observations of them, however, I could not tell the

Freya and Tyr

male from the female. There is no obvious dimorphism between male and female ravens. I had to come to know them as individuals before I could tell them apart, and this task was going to take me significant time and considerable effort.

To begin with, the common raven, or *Corvus corax*, belongs to the class of Aves under the order of Passeriformes in the family of Corvidae. The Passeriformes order of birds is a descriptor depicting their toe arrangement, where three toes point forward and one points back to facilitate the ability to perch. More than half of all bird species fall under this order. A raven is thought to be the heaviest of the passerines, with the largest ones weighing in at as much as 4.4 pounds. Our two friends, however, were a bit smaller, likely more around 2 pounds.

Also, with a bird as ubiquitous as a raven, one would expect to find various subspecies throughout the world, and indeed, this is the case. Many specialists who study ravens recognize at least eight, though others contend there are as many as eleven. While there can be significant genetic differences among raven populations from different regions of the world, there is very little variation in external appearance. The genetic origin of the common raven is from the Old World. It is thought that they entered North America across the Bering Land Bridge during one of the earlier ice ages. As they migrated south and entered different environments, the variations of climate and food sources they adapted to would have caused local evolutionary changes, resulting in the emergence of different subspecies.

Our two ravens, Freya and Tyr, very likely fall into a taxonomy branch called the "California clade," which are ravens found only in the southwestern United States. A clade is a biological group that includes all descendants of one common ancestor. Genetic studies of DNA from ravens across the world indicate that the California clade began to diverge from a common ancestor around two million years ago, likely becoming separated from its relatives during one of the several ice ages. Tests with mitochondrial DNA have indicated that the California clade is more closely related to the Chihuahuan raven—or *Corvus cryptoleucus*, of the American Midwest and Southwest into northern Mexico—than they are to those found in the more northern parts of North America.

Many people don't know the difference between ravens and crows, and numerous people are not even aware that there is a difference. After all, both species are all black. On several occasions I have had to explain to people how to distinguish a raven from a crow:

- First, a raven is noticeably larger than a crow. Although when the bird is by itself this feature is not as obvious, it is quite evident when one sees a raven in the presence of crows. A raven is nominally twenty-five inches long, while a crow's length is around twenty inches.
- Second, a raven flies differently than a crow. A crow's flight involves more-frequent and faster wing flapping, while a raven flaps its wings more slowly and tends to soar more.
- The tail of a raven in flight is wedge shaped, while a crow's tail tends to be straight.
- A raven's wingspan is around four feet, while that of a crow extends nominally three feet.

  * The beak of each bird is another telltale sign. The raven has a larger, curved beak with tufts of feathers at its base on top. A crow has a shorter, flatter, tuftless bill.

  * Finally, the shaggy throat feathers around the raven's neck give the impression of its being thicker.

Once one becomes accustomed to these basic physical differences, one should have no problem with the identification of either species.

A raven's life expectancy is quite long and is among the longest of all the passerines. Although a typical lifespan of a wild raven is between ten and fifteen years, a banded raven has been recorded as having lived twenty-three years and three months—the longest lifespan of any know raven in the wild. But for ravens living in captivity or in a protected environment, that's another matter. Lifespans exceeding forty years are not unusual. The ravens of the Tower of London, who enjoy a very pampered life, have readily lived that long.

Simultaneously while I was learning about their species, I was getting to know Freya and Tyr themselves and coming to learn the personalities of each. They were unique individuals. When perched together, they were physically identical. Differences were undiscernible. But their calls and behaviors were quite different. First, Freya was much more talkative than Tyr. And the timbre of her calls was of a slightly lower tone than that of Tyr. After quite a bit of time listening to both when they are calling together, I came to readily tell who was who. Furthermore, one of Freya's calls was like a warble, which I learned is unique to the female. This communication difference is analogous to what we know from farm animals: a rooster crows, but a hen doesn't. A female raven warbles, but a male doesn't.

The second difference was that Freya was less intimidated of me than was Tyr. She found me curious and allowed me to approach much more closely than did Tyr. He was not necessarily intimidated, but he was much more cautious. One must keep in mind that these were wild ravens with no exposure to domestication. As a result of this behavioral difference, I formed a much closer bond with Freya than with Tyr.

And there was a third characteristic I observed that was different between the two. When they were flying together, Tyr was almost always in the lead. I learned this from their calls. While flying, they were often "chattering," and the one with the lower timbre was usually slightly behind. Furthermore, if I was outdoors when they flew over, I would wave at them and call. When I did that and if I was alone, the second raven would often call back with a single warble! That was Freya! If other people were present, however, she wouldn't do that. Hence, friends were somewhat skeptical of this story. But one day a friend was visiting but standing in a doorway out of sight of the ravens when they flew over. Again I waved and called, and Freya answered with a single warble!

While on the subject of their calls, I would like to expand a bit on raven communication, or at least the communication I have seen occur between Freya and Tyr, as well as between them and me. Ravens are very expressive in their voices and body language. They don't communicate in words as we would understand words, but with their various squawks, knocks, timbre variations, frequency variations between squaws, and so on, they assemble a flow of logic to get across to their intended recipients what they are trying to communicate. And if one is a raven, one understands it very well. Although many ravenologists allege this is not a language, I disagree. Freya and Tyr knew what each was communicating to the other. And I translated several of their calls and knew what they were intending when I heard them. There was a "Where are you?" call when one had lost track of the other. An answer of "I'm here!" would often be the response. Their "Get out of my territory!" was unmistakable. In all, I translated at least a dozen of their calls.

The unique calls and behaviors of these ravens when they perceived a threat from a predator deserve a special mention. Threats from the air in this part of the country are primarily in the form of hawks, specifically the red-tailed hawk. When either raven spotted a hawk, it would take off after it while making a clucking sound. The second raven, hearing that, would join the pursuit while also clucking. One time I witnessed a most extraordinary event. Both ravens launched out of the eucalyptus tree making their clucking sound, flew over a hill out of sight to the west, and in a few moments returned. When they returned, they were escorting a hawk! One raven was flying in front of the hawk while the other was flying behind it—all while making their clucking call. They continued this formation until they reached the eastern edge of their territory, and then they both broke off and turned back, stopped clucking, and flew to the eucalyptus tree. I've seen numerous instances of this type of behavior where they have chased hawks out of their territory, but this particular event was one of the most organized.

Around the time when the ravens first arrived in our neighborhood, we had already been feeding birds from bird feeders we had set up at both the north and south ends of our house at the edge of the canyon. So in order to both welcome our new neighbors and encourage them to stay, we decided to put some food out for them also. There was a path that led past a lone pine tree to the northern bird feeder, which looked like an ideal place to leave something for our new friends. We selected for them a diet of kibbles, which is a dry dog food, and scraps of raw chicken, which they seemed to relish immensely, and twice a day I would leave about a cup of each on the path where they would find it. To make sure they did indeed find it, however, I would call to them. When I got their attention, I would point to where I had left everything. When they saw me doing that, they would normally fly over to one of the lower branches of the pine tree and prepare to hop down for breakfast or dinner. Over time this developed into a little game.

Because Freya was much more talkative than Tyr, since shortly after their arrival she and I had been "communicating." I would often call to her with a certain tonal inflection that I would repeat three or four times, and she would answer with a warble. I used a very similar call when calling her for food. Even when she and Tyr were far away up the canyon, if she heard me, she would often warble back and fly down to the eucalyptus tree. Very often when she was flying overhead, I would wave to her and call, and she would warble back. Occasionally throughout the day when I was home, I would hear her warble. Whenever I did, I would usually step out onto the balcony and call back. Any given week, this would happen quite often. We would call to each other for several minutes at any one time. This too was like a little game!

A most extraordinary event happened one day. That afternoon I had put some kibbles and chicken scraps down for the ravens in the traditional spot by the pine tree, as I usually did. About an hour later, I looked out and saw a coyote eating some of the kibbles. I picked up the binoculars, and while standing out of sight, I watched the coyote through the window for a few minutes. While I was watching the coyote, Freya flew over to the lowest branch of the pine tree and started scolding the coyote. This was a similar call to that which she and Tyr would make when trying to drive off a bobcat. The coyote completely ignored her and continued eating. She was apparently hungry, and the coyote was eating food that was unmistakably hers. She clearly was frustrated. And then to my complete surprise, she started to warble. Her warble was her "I'm happy!" call, which she definitely was not at the moment. I realized she was calling me to come out and chase the coyote away. I stepped out onto the balcony, shooed the coyote, talked to her, and then went back inside to watch her reaction. She immediately gave the "Food!" call to Tyr, which sounded something like a squawking quack, then flew down to the kibbles and started eating. Tyr flew over to join her.

So, why would I think her warbling was a call for me to intercede? She was perched in the pine tree trying to discourage a coyote that was ignoring her. From her past experience, she knew that if she warbled, then I might show up. And if I did show up, the coyote would leave. Maybe it was my imagination, but I thought this action was intentional and required rational thought!

And so began our unique relationship with our new feathered neighbors with whom we were destined to share a myriad of exceptional experiences.

CHAPTER 5

Family Life in Ravenworld

● ● ●

FOR SEVERAL YEARS I STUDIED the behavior patterns of Freya and Tyr and observed the various cycles of their lives. For many seasons they would nest somewhere at the end of a short canyon north of our house. The terrain, however, was too rugged and steep for me to attempt to find out exactly where. But in the late spring, they would fly out with that year's hatchlings to begin training them for all the things they would need to know to survive in the world of ravens.

We saw no little ones the year of the West Nile Virus outbreak. I strongly suspect that hatchlings were born that year but were infected by the virus very early and died in the nest. This suspicion was acceptably confirmed by the fact that the following year the ravens built their nest on the far eastern edge of Snover Canyon, a long distance from the location where they had earlier nested for so long. However, this new nesting location was short-lived. In August 2009 there was a major brush fire that encompassed their entire territory and left any resemblance of that year's nesting site in ashes. The following year they would have to relocate, but to where?

Late February to early March is when our ravens usually started building their nests. Because of the fire having destroyed their previous nesting site, we were curious as to the location they would select in 2010 to build a new one. To our delight, after a day or two of observing their flying about with twigs, I learned that they decided to build it in one of the eucalyptus trees immediately across from the balcony, directly east of our house—a distance less than fifty or sixty feet away as the raven flies!

Our house sits at the periphery of a fairly steep embankment that descends nominally seventy or eighty feet to the edge of a dry creek, or arroyo. Located at the house side of its threshold is a copse of a half dozen or so eucalyptus trees configured in three distinct clusters. The ravens selected as their nest site the southernmost tree in the center cluster. It was not the tallest tree, but it did provide considerable shelter within its new limb and leaf growth. Good fortune smiled again when we learned that the new nest position was to be on our side of the tree. This was an ideal location that would enable us to observe their entire process of nest building, egg laying, hatching the little ones, and caring for the hatchlings until it was time for their flight training.

It was during the first week of March of the year 2010 when I noticed the ravens flying about the eucalyptus trees with an apparent urgency. By midweek I began to pay closer attention and decided that this was going to be the year that I would keep a day-to-day journal of their nesting events. I figured that I may never get another opportunity like this one. I kept this daily log for nearly six months,

Copse of eucalyptus trees

and at times the rapid daily development of the youngsters seemed to progress like a scene in a time-lapse video.

So much happened within such a short period of time when I was keeping notes that it seemed like the best way to tell this part of their story was to summarize the daily activities of our raven family. The chapters that follow are the daily summaries that I recorded that year.

CHAPTER 6

Preparing for the Blessed Event

● ● ●

MARCH 11, 2010: TODAY, THERE was a flurry of activity with both ravens flying back and forth to the eucalyptus trees carrying twigs of various shapes. They obviously were building a nest in one of them. So I followed them closely for about twenty minutes and noticed that they were flying into the southernmost eucalyptus tree of the center cluster near its top, among new growth. That was the chosen site this year where they would be building their nest!

I watched off and on for several hours as they took twigs and carefully placed them in just the right positions. The twigs were all of about the same size, as they had cut them (bitten them off) to that length.

The nest was ideally suited for an observer. With binoculars from our upstairs window, I could look down into the nest and see the majority of the bird. When the adult raven was off the nest, I could not see into it completely. But I could see well enough to detect any major developments. This location was going to be ideal for watching the family progress!

March 13: Much more active nest building took place today. The nest format is now becoming obvious. As an experiment, I set up a Celestron telescope to look into the nest, but the telescope had a very narrow field of view and was difficult to position. A slight movement by the raven required it to be reoriented. Nonetheless, I decided to keep it in place, as it would still be handy for spotting the eggs and hatchlings.

March 15–16: More nest building occurred off and on throughout the day. The nest was taking shape so sufficiently well that it now became much easier to locate where the branches forked in the tree, as the twigs for the nest were denser and more obvious. Both ravens were busy chasing the hawks away, and it appeared that now they were much more serious in doing so!

Nest building continued into the following day, but by late morning it seemed to have definitely slowed. It would appear that except for the fine-tuning and interior decorating, the job was fairly well completed. By midday, there was no activity at all.

March 17–18: Nest refinement picked up a bit today. One raven would bring material to the second, who was on the nest and assembling it. The courier raven would fly over with material in its beak, pass it off to the one in the nest, and then fly away for more. I suspect it was Freya on the nest, packing the material in just the right spots, as she would nestle down into her new construction and squirm around as if to pack everything around the bottom with her chest feathers. This afternoon she was singing her "all is right with the world" song. She was happy.

More nest fine-tuning took place the following day, but there seemed to be a little less activity dedicated to it. Both ravens were out flying around together for most of the day. I took several photographs of one of the ravens standing on the nest.

March 19: Some final touches were put on the nest today, although it looked like the general activity was diminishing. One of the ravens—and I strongly suspect it was Freya—got into the nest and buried its chest into the bottom while ruffling its feathers and wiggling back and forth. Eggs should appear any day now.

March 20: It was exactly a year ago today, the first day of spring, that Freya settled into her nest for the duration of the egg hatching. I had watched their activity across the canyon all last year, and on that specific day there suddenly was only one raven flying about. Although eggs will likely not be showing up on this day this year, they should be arriving very soon. In fact, the nest looks to be fairly well finished.

March 21: Again, not much activity occurred involving nest refinement. Both ravens flew to it only a couple of times today that I noticed, but neither dwelled there very long. And the nest does seem to be completed. They have lined it with soft material, including what looks like their own feathers. Meanwhile, both are flying around, seeming to be very happy.

March 23: I think the big day has arrived, as Freya was on the nest for the majority of the day today. It appeared that she has indeed laid some eggs. She had flown over to the pine tree in the morning for food, but shortly thereafter had gone to the nest, where she stayed for most of the day. She did leave the nest in the evening, however. But before she left, it appeared that Tyr had come by with some food. Once a raven has laid eggs, the female stays on the nest incubating them until they are hatched, and she relies upon the male to bring her food. Furthermore, according to others who have studied ravens, the male raven does not help with the incubating.

March 24: Freya was already on the nest first thing this morning and didn't come by the pine tree for food. She was serious this time and had eggs to incubate! I watched her throughout the day, and she

Freya on the nest

did not leave. Tyr has been around all day, and hopefully he has kept her fed.

Early in the evening, she left the nest for only a few minutes but came right back. She didn't seem to go anywhere in particular. I suspect she just needed some exercise.

March 25: Again, Freya was on the nest all day. She seemed to be very alert, as she was constantly looking around. Tyr was nearby for most of the day. He picked up some of the raw chicken I left for them this morning, and I trust he took some to his mate on the nest.

At 5:25 p.m. Freya flew off the nest and was gone for about a minute. Like yesterday, she didn't go anywhere in particular; she just glided away and came back. I will watch for this behavior more closely to determine if it's typical.

March 26: Freya was on the nest all day. She would change positions occasionally and appeared to rotate the eggs. At 3:25 p.m. she left the nest for about eight minutes. She was probably taking a break, getting a little exercise, and possibly getting a drink of water, as I saw her fly up the dry creek. There was a trickle of water still coming down the upper part of the dry creek before the water disappeared altogether. Unfortunately, the nest is just deep enough that I could not see the eggs when she was off it. But its positioning for virtually all other observations is nearly ideal. Freya returned to the nest and settled in for the remainder of the evening and presumably throughout the night. We have a full moon coming up, so I may be able to spot her at night with the binoculars.

March 27: For most of the day, Freya stayed the course doing her duty sitting on the nest. But an intriguing event happened around six in the evening. She had come off the nest around that time and had flown over to the pine tree. She was already at the pine tree by the time I spotted her. I know it was she because when she saw me, she warbled. Furthermore, I was able to walk over to the trail with some chicken scraps while calling her, and she never moved. Tyr would have flown away. But shortly after I had left the food and gone back into the house, Penny told me that there was a bird on the nest! Either Tyr substituted for Freya for a short time, or Freya was awfully fast! This was exciting, as it implied that contrary to earlier information I had obtained about raven nesting habits, maybe the male does indeed occasionally fill in for the female in egg-warming duties. We are definitely going to have to be especially observant for this possible exchange from now on!

March 29: Today I made a truly exciting observation. Around three o'clock, I was on the balcony watching the ravens. One was on the nest, while the second was flying around locally. (Notice that I said "one" and not "she" was on the nest.) I watched the one flying around, as it stayed close by. It went down for a drink to the trickle of water still running in the upper area of the normally "dry" creek. I watched it drink. It then flew up to the large eucalyptus, called a few times, but then gave out a unique warbling call—a call that only the female raven makes! In addition to that specific call, this raven was also very "talkative," a characteristic that Freya demonstrates, while Tyr is fairly quiet.

At this point I had been presented with a most exciting dichotomy—either both the male and the female can create this warbling sound, which ravenologists claim is an ability of only the female, or the male may occasionally give the female a break and sit on the eggs, which is thought to be a realm of only the female. Again, based on what I have learned from my research on ravens, ravenologists allege that throughout the egg incubation period, the female stays on the nest with the eggs all the time, and the male brings food to her. This implication is that she does not leave the nest, nor does the male take turns with her and occasionally sit on the eggs. However, during the course of my observations, I have seen Freya leave the nest for a few minutes in the early evening when she gets a drink of water and fly around for a while. So at least part of this assertion is not rigorously the case.

Meanwhile, Penny joined me on the balcony to lend a second pair of eyes to this event. While I was still observing the raven that was on the nest, the raven that was "out and about" then flew to the broken limb of the tall eucalyptus and landed where I could see both it and the nest in the other tree. Both birds gave short "hoots" to each other that sounded like a cross between the sounds of a dove and an owl. It was clear that both were talking, as the voice timbre was different for each. The second raven then flew from the broken limb over to a branch adjacent to the nest on the other eucalyptus. At this point, the raven that was on the nest got up, stepped off it, and flew over to the nearby pine tree. The bird on the adjacent perch near the nest followed but landed on a different branch in the pine. By having two people watch this, we were able to keep track of the actual location of each individual bird. At this point, the bird that was originally on the adjacent branch in the nest-bearing eucalyptus flew from the pine tree directly to the nest and got on it. The raven originally on the nest flew to the ground where there were some food scraps. These ravens had undeniably switched egg-warming duties! Freya was the one we had originally seen flying around, while Tyr had been sitting on the nest! We will definitely be watching for a repeat of this behavior over the next couple of weeks!

This exchange was a most fascinating observation, and it explains the apparent paradox we observed two days ago. Egg-warming duties are at least partially shared between this specific mated pair of ravens!

March 30: Once again we witnessed a change of nest-sitting duty! Just before noon, as I was looking with binoculars at the raven in the nest, I watched that raven stand up, step to the edge of the nest, and then fly over to the nearby pine tree. Meanwhile, during this time the second raven was standing at the end of the broken branch of the large eucalyptus tree. It never left its perch when the other one flew off. Instead, it did a few bill wipes and just stood at the end of the broken branch for about a minute. It then left the branch and flew to the nest, where it got in and nestled down onto the eggs. The raven originally on the nest had in the meantime flown to the ground, where it was pecking at some kibbles that had been left out for them. A "changing of the guard" had clearly and unambiguously occurred!

From now on we will pay closer attention to the raven that is flying around to identify whether it is Tyr or Freya. Their personalities are quite different, so for the most part we should be able to identify who's who.

Around six o'clock, I think there was another change of the guard. When I looked out the window, the nest was already empty, but one raven was perched nearby. Meanwhile, the other raven was flying down to the stream, where it stopped to get a drink of water. It then flew over to the pine tree and down to the ground to peck at some kibbles. About this time, the raven that was perched nearby flew over to the nest and nestled down in it. What I believe happened is the one that got the drink of water had just come off the nest, while the one perched nearby was waiting its turn. Furthermore, I noticed that the one perched nearby had a somewhat duller sheen to its plumage than the other had. I think that raven was Freya, as I have noticed a slight dulling of Freya's plumage over the past few days. What I believe happened is Tyr had just left the nest for a break, and Freya was standing by to take over sitting duty.

March 31: Today, at 6:03 p.m., when I looked out the window, I noted that there was no raven on the nest. I went outside to locate it and found that it was sitting on the nearby century cactus. I then saw our next-door neighbor. We talked about the ravens' recent habits and decided that we would wait for one of them to return to the nest, as they have never been off for longer than eight minutes before. While we were waiting, Freya flew by a couple of times but didn't stop by the nest. We wondered if something may have possibly spooked both of them, as I also never really saw Tyr. Our neighbor suggested that it might be his (the neighbor's) presence, so he left.

Within about two minutes after he left, Freya came back to the nest. All in all, she was off the nest for at least thirty-three minutes, which is the longest we have ever seen the nest unoccupied. Today was a fairly cool day also, as the temperature was around 49 degrees Fahrenheit while we were watching.

April 2–3: Around four o'clock today we heard a cacophony of squawking from the ravens—both of them. I went outside to investigate, and two vultures were flying very low over the house and eucalyptus trees. Both ravens were out trying to chase them off. While I was out there, the vultures made a couple more low and slow passes, and then flew off to the southeast. Both ravens, meanwhile, were still excited and flew up and down the canyon while calling out. Most of the calling was being done by Freya, who tends to have a lot more to say.

As they came to realize that the vultures were no longer a threat, they started to relax. Freya did a few aerobatics and landed at the very top of the pine at the west end of the driveway. She then gave her "Here I am" call to her mate. I spotted her, waved, and called back to her. At this time, Tyr flew over to her, and she left her perch. Together, they flew back toward the house, calling all the way—Freya doing most of the calling. I stood on the porch waving to them. As they flew over me and when they were directly overhead, Freya gave out a single warble call as if to say hello. She then went back to her other calls. I watched them for a couple more minutes, and then one of them—I suspect it was Freya—went back to the nest. The nest was vacant for probably six or seven minutes during all this.

I have suspected for a couple of years now that Freya actually recognizes me. Maybe it's my imagination, but I really think that warble was an intentional greeting.

Well, today turned out to be a major day for raven watching! At 4:20 p.m. I went outside to put out some birdseed for the other birds, as well as some kibbles and raw chicken scraps for the ravens. About twenty minutes later, after I had returned to the house, I saw Tyr in the eucalyptus tree preening his feathers. I watched him for a short time, and then he flew over to the pine tree near where the kibbles and chicken had been placed. After a properly cautious approach, he flew down to the food source and ate a couple pieces of chicken. When he finished, he picked up a large piece, flew over to the nest, and gave it to Freya!

Since he did this once, I thought he might repeat the process, so I watched him more closely. Indeed, he did. He made the same cautious approach to the food, ate a couple of pieces, picked up a large piece, and flew over to the nest. I had the binoculars on him during this entire transfer and could see Freya's beak moving.

In anticipation that he would go back a third time, I aimed the Celestron telescope directly onto Freya's head in the nest. While I was doing that, Tyr had already gone back to the kibbles and chicken and was making his final selections. As he flew back to the nest, I looked through the telescope eyepiece and saw him arrive and perch on the nest's edge. I had a clear view of both of their heads! She opened her beak like a baby bird, and he put the food into her mouth. He did two or three movements that looked like they might be regurgitations and then flew off. She moved her beak as if she were swallowing. He repeated this food delivery at least two more times. I was a clear witness of at least three of his feedings of Freya!

As I've said before, ravenologists have long claimed that the female sits on the nest, and the male brings food to her. We now have seen not only that occurring, but we have witnessed the female also taking a break at least once a day for a few minutes, and most importantly, we have observed the male taking turns sitting on the eggs!

Around ten o'clock the following morning, I witnessed another feeding of Freya on the nest. I watched Tyr pick up some of the chicken scraps, fly around the back of the eucalyptus trees, land momentarily in a dead

Tyr feeding Freya

tree in the arroyo, and then fly directly to the nest, where he gave the food to Freya.

April 5: It was a rainy day today, but all day Freya stayed hunkered down on the nest. The weather started lifting and began clearing around midafternoon, and around five o'clock. I took out some kibbles and chicken for them. While I was putting birdseed out for the other birds, Tyr was in the south end of the canyon perched on a dead tree that had been severely damaged by the fire. I called to him and waved. With that, he called back, left his perch, and flew directly to the pine tree where the kibbles were. I kept an eye on him while he picked up several chicken pieces and flew over to the nest to feed his mate. He repeated this several times. Because the weather was bad today, food was probably scarce, and they both were hungry.

At 6:40 p.m. I looked out the window at the nest and saw that it was empty. Freya had left it and was out getting a little exercise. I went outside to investigate. Both ravens were cruising up and down the canyon, catching the updrafts and calling. They played in the sky like this for about another five minutes before returning to the eucalyptus tree. There seemed to be a bit of reluctance for either to fly directly to the nest, so I went back inside. Once I did that, Freya immediately returned to her nest duty.

April 6: Again this evening around six, Freya left the nest for a little exercise, and as before, she didn't stay off very long. Shortly after she went back to the nest, Tyr went over to our kibble source and brought back some food for her. I had suspected this was what he was going to do, so this time I had a camera ready—and I got a couple of good photos of him feeding her!

If March 23 was indeed the date of the first eggs laid, then today should be about two-thirds of the way through the incubation period. Hatchlings should begin arriving around April 13, so we'll see. Several times throughout the past week we have seen Freya turning the eggs over. She gets into the nest and with her beak moves the eggs around before settling down upon them.

April 7: This evening around seven, both ravens were perched together on the broken limb of the eucalyptus tree, and no one was on the nest. Unfortunately, I didn't see which one came off the nest nor exactly when. But during the interval I was watching them, one of them—who I believe was Freya— left the limb and flew back to the nest, while the other flew over to the pine tree where the kibbles were.

April 8: A most noteworthy event occurred today. A little after four this afternoon, I heard Freya warble outside. That meant she was not on the nest, as she has never warbled while sitting. I picked up the binoculars and went outside to see if I could find her. However, "someone" was sitting on the nest. Then I saw a raven flying low over the dry creek where a damp trickle was still coming out. It then came back and did two or three slow swoops close to where I was standing on the balcony. I could not identify which raven it was. I called and waved to it, and it swooped back even closer and then flew over to the pine tree. But unlike the pattern Tyr usually makes, this raven not only flew slowly and very close to my location, but it was occasionally "talking." When it flew to the pine tree, it went directly to the outer limb overlooking where the kibbles are normally put, instead of landing in the opposite part of the tree and slowly working its way to that limb. Furthermore, this raven's plumage was a little duller than the one that was presently sitting on the nest. Based on all of this, I am certain this raven was Freya!

But what happened next was most fascinating. From her perch on the outer limb of the pine tree, she leaped off and did a slow and low swoop directly over to the nest and perched on its edge, while I watched all this with binoculars. Once she was on the edge, Tyr stood up and stepped off. Freya then stepped on and nestled down over the eggs.

Unfortunately, I didn't know when Tyr first got onto the nest, as that would have been most noteworthy to learn. Throughout all of these raven observations I have not been able to determine how long the male raven actually sits on the eggs. But at least for selected periods, I know for certain he clearly does!

April 9: Today produced another most remarkable raven event. A little before three thirty, I heard both ravens calling. With both calling, that meant that Freya was off the nest, so I picked up the binoculars and went outside. There was a bit of a breeze this afternoon, and they both like to catch the updrafts and soar with the wind. When I was out on the balcony, I saw them at a fairly high altitude up in the far canyon. I watched them soar for a few seconds until they turned to head back. At that point I waved and called to them. While I was waving and calling, Freya broke formation and swooped down. While she was descending, she did a small aerobatic action where she collapsed her wings and made a short dive. But what she did next was wholly unexpected.

While I was leaning on the balcony railing waving at her, she swooped by no more than ten feet away and did an aerial loop! She went full circle. At first she ducked down to gain speed, and then she pulled up and went right over the top, upside down! I have never seen that maneuver performed by a bird. On numerous occasions I have seen ravens—in fact, this specific raven—do a 360-degree barrel roll. But I have never seen her perform a "loop"!

She then joined Tyr as he flew by. They both did several passes over the house, but Freya flew much lower and much slower as she soared by where I was standing. She swooped past twice before they both landed at the top of the eucalyptus tree. Once perched there, she called out with her warble call. I waved at her and called back. She then left her perch, glided over to the nest, and settled in. This whole action was completed in less than ten minutes.

Around five o'clock this evening, she left the nest again for only a few minutes and then went directly back.

April 11: Today was cool and overcast (the temperature was around 49 degrees Fahrenheit) with about a ten-mile-per-hour wind, and Tyr stayed around the eucalyptus trees most of the day. When I checked the nest midmorning, Freya was rotating the eggs with her bill. Once she was satisfied everything was properly positioned, she fluffed up her feathers, spread out her wings, and settled down to continue warming the eggs.

A little before three in the afternoon, I heard both ravens calling, so I again went outside. Freya was off the nest; in fact, she was on the roof. As I stepped onto the balcony, she flew off and did a series of very slow flight maneuvers, as if she were playing in the wind. She flew very close to where I was standing, and then she caught the wind and went up the canyon. All the time she was performing these antics, Tyr stayed on his perch near the top of the large eucalyptus tree. When Freya returned, she swooped around slowly and joined him. I called to them both, and she warbled back with that unique call of hers. She rested there for a few moments and then glided back to the nest. She called out a few times as she was settling down on the nest. Normally, she remains silent at this point.

About an hour later, she left the nest again for a few minutes. She flew around for a short time, made a few calls, and then returned.

April 13: Ever since the first time I observed that Tyr will occasionally sit on the eggs, I'd wondered how long he might actually stay on the nest. From an observation made today, I may have at least one data point.

Around one o'clock I noticed that no one was on the nest, and both ravens were having fun catching the wind currents from the breeze that was blowing. The day was cool, the skies were partly cloudy, and the winds were gusting lightly over the canyon. One raven, which most likely was Freya, was soaring in the wind and calling out fairly regularly, as she usually does. Since it was now getting fairly close to the time when the eggs should be hatching, I went out onto the balcony with binoculars to look into the nest to see if I could determine if there was any activity. I observed none.

Meanwhile, one of the ravens came back and landed near the top of the large eucalyptus tree. It then flew down as if it were going to land in the nest but instead continued on to the adjoining tree. This behavior is more typical of Tyr, as Freya would have gone directly to the nest. Tyr, I believe, is more cautious about giving away the location of the nest when I'm around, whereas Freya knows me and trusts me. It then left that tree and went to a perch on a flower-stem outgrowth of the century cactus, a place where Tyr often goes but where I have rarely, if ever, seen Freya. From that perch, it then glided over to the nest. I am virtually certain that it was Tyr who got on the nest.

Meanwhile, at approximately two this afternoon, I heard one of the ravens outside calling, and shortly thereafter I went outside to see what was motivating its interest. One raven was casually flying around and calling, while the other was on the nest. I went inside to look through the telescope to observe the one on the nest in more detail. It seemed to be somewhat nervous, as it was pecking at something in the nest. I was wondering if maybe a hatching was taking place. But while I was studying that in detail, the view suddenly filled with the second raven arriving at the nest. As this one landed on the rim of the nest, the one presently there immediately got up and flew off. The new arrival then stepped in, settled down comfortably, and was much more relaxed and at ease. The new arrival had to be Freya. First, the sheen of its feathers seemed a little duller, and second, when it was flying around, it was "talking"—a trait characteristic of Freya and virtually absent with Tyr.

With this "change of the guard," coupled with the earlier observation of the typical behavior of Tyr, I believe I witnessed a period of about an hour where the male had been sitting on the nest!

April 14: At 11:18 a.m. I noticed that the nest was empty and went outside to investigate where the ravens were. They were out soaring together. I called to them, and they returned to the tall eucalyptus tree. I continued calling while they were in the tree, and Freya called back with her warble. Once she did that, I then knew which was which and followed her actions thereafter. It was definitely she who returned to the nest.

Again about an hour later I heard the ravens calling, so I once more checked the nest. Freya had left it, and both ravens were having fun in the sky flying about. I could hear Freya "talking" virtually continuously, as she normally does. Then one of them returned and perched on the century-cactus shoot, while the other was still flying—and calling—in the background. The raven on the cactus then flew up to the nest. Although I was not certain, I strongly suspected that this raven was Tyr.

Approximately a half hour later, I watched another change of the guard. By chance I was looking out the window when the "free" raven flew over to the nest and landed on its rim. The raven on the nest stood up and flew away. The new arrival then stepped onto the nest and settled down. So it was quite possible that the "replaced" raven was Tyr, as I had suspected.

The nest was empty again a little after five in the evening, when the ravens were out flying together. Again I called and waved to them. They both landed in the eucalyptus tree, and Freya warbled back as I called. She then returned to the nest. As before, I knew this was Freya, as I had followed where she went after she warbled.

The ravens left the nest today much more frequently than I have seen so far since logging their behavior. Also, the eggs should be hatching very soon. Maybe this behavior is indicative of the eggs being in the process of hatching.

CHAPTER 7

So Many Little Beaks to Feed!

● ● ●

APRIL 17: WE HAVE HATCHLINGS! Or at least we have very strong indirect evidence of this being the case. There has been a flurry of activity with both parents going on and off the nest this morning. At 8:06 a.m., I saw a "change of the guard" as one flew up to the nest and replaced the one already there. Later that morning, at 10:24 a.m., I saw both entering and leaving the nest. When either raven entered the nest, it would hunch over and seem to be feeding something, as one would expect a parent bird to feed a baby bird. The nest was sufficiently deep that I couldn't see the hatchlings, but all evidence pointed to it or they being there. Once the feeding was complete, Freya would settle down onto the nest, but she would approach very carefully. It was as if there may be one or more eggs still remaining and she didn't want to injure the newborn. About an hour later, both parents were involved in this feeding behavior. On one occasion, one raven stood in the nest and squatted while spreading out its wings as if it were trying to keep the hatchling warm. Meanwhile, the feeding activity kept up for several minutes.

All afternoon there was continuous activity, with both parents buzzing in and around the nest. Both birds changed places often. Although the nest was uncovered more frequently than it has been to date, it was never vacated for very long. There was at least one adult attending it more or less continuously—sometimes sitting fully on the nest and sometimes squatting over it with wings outspread. But all throughout this nest activity, the adults were lowering their heads into the nest as if "feeding" someone.

Around midafternoon, Freya went out for a long flight over the canyon. And while she was flying, she called out with her warbling call several times. That call essentially says, "I'm queen of the mountain, and all is right with the world!" It was obvious she was happy. Maybe she was also saying, "My part of the job is done. Now it's up to both of us!"

Since the hatchlings grow very fast, soon little heads should start popping up over the nest rim and we will be able to count how many we actually have.

April 18: Today's activity was very much like that of yesterday, with a strong sense of urgency as both parents flew on and off the nest. During much of the time I observed,

The first little hatchling

however, one parent or the other was actually sitting on the nest. But throughout that sitting activity, that adult would occasionally feed someone. It would poke its bill down toward the bottom of the nest and act like it was regurgitating food for a little one. Although I have not actually seen any of the little birds yet, there is clearly no doubt of there being one or more little hatchlings present!

When I put out some chicken scraps around midafternoon, I noticed Freya was out flying, so I called to her. She saw me and swooped overhead low and slow several times, and then she landed in the eucalyptus tree. But rather than coming down for something to eat, she flew back to the nest.

Around six fifteen, after the sun had slipped behind the horizon and the glare was off the nest, I was watching one of the adults through the telescope (very likely Freya) feed the hatchlings out of sight below the rim of the nest when one little beak poked up above the rim. It was a very light-yellow beak that was wide open and saying, "Feed me!" I had now verified that there was at least one hatchling!

I checked back a few minutes later. There was another feeding going on, and I saw two little beaks poke up—this time more clearly. They were yellow to light orange, and a little darker around the outside of the beak. So now we knew there were at least two hatchlings—and possibly three!

April 19: Today both adults were on and off the nest, bringing food to the little ones. I still do not yet have a final count as to how many there were, and it likely will be another week or so before I know. The little ones are still very weak and unable to hold themselves up. But as they get stronger, their little beaks will

More little hatchlings!

poke up higher and stay above the nest rim longer. Then I will be able to make the final count!

April 20: More care and feeding of little hatchlings took place today. The day was cool and overcast in the morning, and by midafternoon it had started to rain. At that point Freya settled over the hatchlings and spread her wings like an umbrella to keep the new brood warm and dry. I still don't know how extensive her little brood is.

April 22: It was cool again today and remained a little drizzly in the morning. For most of the day, Freya stayed hunkered down in the nest, keeping the little ones warm and dry. She did go out for some flight exercise in the afternoon, however. Again, I'm still trying to learn how many hatchlings there are.

April 23: Today's activity was very similar to that of the past few days since the hatchlings arrived. The ravens were rapidly shuttling back and forth, feeding the little ones, and Freya often lingered on the nest to keep the hatchlings warm.

Our neighbor lent me his camera and tripod, which I set up and then used to take a few pictures of the nest—both when it was empty and when one of the parents was present.

April 24: Normal activity prevailed on and off the nest throughout the day. Toward evening, when the sun angle was such that I could see much better into the nest, I was able to glimpse little orange beaks popping up over the rim. With a combination of the binoculars and the telescope, I confirmed that there were at least three little ravens with a possibility of a fourth. One was definitely larger than the other two. In fact, all three looked to be a day or two apart in age, as if the eggs were laid within that span of time. The biggest one was poking its little beak up fairly vigorously, and the other two were attempting the same, but with progressively descending strengths. Although I could not see over the rim directly into the nest, I did sense that there may be a fourth hatchling present, as I thought I saw movement. But if there is a fourth, the little one is still a little too weak to stand up on its own.

April 25: The hatchlings are becoming a little more active, as they are getting stronger by the day. When the parents are off the nest, I can readily see one or more little heads bobbing above the rim.

Early this afternoon, Freya was in the large eucalyptus tree, calling. As I went outside to see what she was calling about, she flew over to the pine tree where I had already left some pieces of chicken for them. She apparently was looking for food. Everything was gone from this morning, so I went into the freezer to get some more chicken to cut up and take out. I spread out a handful of chopped chicken and came back in. By the time I had looked out the window again, she had already found them and was on the ground picking up pieces. Once she had a beak full, she immediately flew over to the nest to feed the little ones. I then went back to whatever it was I was doing.

A few minutes later, I heard her squawking from the pine tree. It was a somewhat unusual call. I looked out and saw a bobcat walking down the trail away from where I had placed the chicken. Freya was very annoyed that the bobcat was stealing her private supply! She darted at it a few times and then landed on the balcony railing, looking down at it. I grabbed the camera and stepped out onto the balcony near where she was. She did not fly away! Instead, she squawked while we both looked for the bobcat. I came back into the house and went outside through the garage to see if I could catch up with the bobcat. By then both ravens were flying low over the dry creek and squawking their "Bobcat! Predator!" call. They had found the bobcat. I watched for a few minutes, as they seemed to be concentrating, and sure enough the bobcat came into view for a few moments before disappearing again. Very little gets past these ravens!

This evening when the sun went behind the hill, I could definitely see that there were three little orange beaks poking up. But there also undeniably seemed to be a fourth beak present as well. In a few days, when they all are stronger, we should know for sure!

April 26: Around three in the afternoon, Freya was calling from the eucalyptus tree, so I went outside to investigate. I noticed all the food I'd put out was gone, so I thought maybe she was calling about that. I took out some chicken pieces, along with a few kibbles, and spread everything out on the ground near the pine tree where I usually do. I then called to the ravens and went back into the house, and sure enough, they both flew over. I went out onto the balcony to watch them as they both made trips back and forth to the nest to feed the hungry little beaks.

While I was sitting out on the balcony, a silver gray fox came down the path and headed for the spot where the ravens were still picking up scraps of chicken. They saw it and flew to the pine tree. I then stood up and tried to discourage the fox from eating the ravens' food by waving at it. It acknowledged me but then went back to eating. So I went downstairs and outside, and walked over to where the fox was, all the time talking to it. The ravens, meanwhile, were still in the pine tree, with Freya perched on a low branch closest to the path. As I approached the pine tree, Tyr flew off, but Freya remained. I passed the pine tree within a few feet of where she was perched, but she knew I was focused on the fox and of no threat to her. I shooed the fox away. When I returned, she was still there. This time I stopped and talked to her, but this time she flew off. As I moved away and called to her, however, she came back. And now with the fox gone, she flew down to the chicken pieces.

For maybe twenty minutes or so I watched both parents continue their rounds back and forth, feeding the hatchlings until nearly all of the chicken pieces were gone. But now the fox, thinking the coast was clear, came back. This time I let him eat, as the ravens were fairly well finished.

Later in the evening, with the sun farther in the west, I could see into the nest much easier, and behold! It was confirmed! We definitely have four hatchlings! All are now strong enough to poke their little heads up, and I was able to count four little orange beaks! I even got a photograph of them. It is amazing how quickly they are building their strength, as each day they are getting noticeably stronger. This afternoon when the parents were feeding them, I heard one of them squawking weakly. For each group of hatchlings that Freya has had over the past few years, there has invariably been one little raven that is especially noisy. So it sounds like there will be no exception this time!

April 27: I had a most enlightening experience with the ravens this morning that seemed to have brought our relationship to a new level. I was a little later than usual taking their ration of chicken and kibbles out to them. When I arrived, I looked around for them, but they were nowhere in sight. I called out for them and heard a faint call in return from the distance.

I called again and also got a response, so I assumed that they knew that "breakfast" would be available fairly soon. When I finished chopping a few pieces of chicken, I called again and heard the response immediately behind me. Both ravens had flown into the pine tree while I was cutting the chicken, and I hadn't seen them. They were waiting very patiently.

With a small container of chicken pieces and one of kibbles, I slowly walked over to the spot under the pine tree where I usually put their food, while talking to them along the way. Both stayed where they were—especially Tyr, who has typically been so skittish. I spread the food out while trying to keep my back to them so they would not perceive me as a threat. I continued talking to them and slowly backed away. I was no more than twenty feet away when Freya flew down to pick up a meal. Tyr was more cautious, but he didn't immediately fly away as he has typically done. Shortly thereafter, he too flew down to the food. It seemed that he was now starting to trust me more!

April 28: There is not much new to report today, as both parents were shuttling back and forth, bringing food to the little ones. They did seem to stay off the nest for longer periods, however. And when the parents were not on the nest, I noticed that the little birds were generally asleep.

April 29: Today's raven activity was essentially the same as that of the past couple of days, although the weather was windier, and both parents spent much more time catching updrafts and playing in the air. Both parents also made frequent trips to bring food to the nest. I gave them some additional chicken pieces so that they would not run out of food to take to the hatchlings. Whenever I do that, I always call to them, and they know exactly what that call means—no matter where in the canyon they may be!

The little birds seem to be growing fast. Their little bodies are now fairly well covered with pinfeathers, and flight feathers have started to grow from the trailing edges of their little wings. They have been sleeping a lot, and while they are asleep, their little bodies are developing. But when they're hungry, their little orange bills pop up very enthusiastically! Mom's usually the first to respond.

April 30: There was not much new going on in Ravenworld today. The adults seemed to be staying off the nest for longer periods, letting the little ones get their much needed rest. They also seemed to

Feed me!

be feeding the little ones on a fairly fixed schedule but not continuously. Many times the hatchlings were moving about the nest with their "giant" orange beaks wide open, saying, "Feed me!"

But the adults tended to ignore this action if it didn't occur at the correct times. I have tried to note when the regular feeding times occur. There definitely is one around six o'clock in the evenings that I have observed so far.

This evening when I chopped up some chicken and called for the ravens, both Freya and Tyr flew over to the pine tree. Tyr is apparently getting accustomed to me and learning to trust me more. They both stayed in the tree, with Freya closer and Tyr farther away, as I took the chicken scraps over to them.

May 1: The little guys are growing as we watch them! The older ones are now fairly well covered with gray "fuzz," and their little pinfeathers are becoming darker. The youngest one is starting to develop its pinfeathers, but it still has much exposed skin.

The adults started feeding them at 5:00 p.m. today. Those little orange beaks were probing the air very enthusiastically! Although it was very difficult to tell through either the binoculars or the telescope, it seemed like the oldest one had its eyes open, while it appeared that the eyes were still closed for the remaining younger ones.

While watching the adults make their feeding trips this afternoon, I observed another fascinating phenomenon. One time when one of the adults flew to the nest, it didn't feed the little ones. Instead, it cleaned up the nest! It pecked at bits of debris the little ones had left and flew away. I've never heard of ravens—or any bird for that matter—doing that.

May 2: Today the adults seemed to be taking a kind of laissez-faire attitude toward the little ones. It was almost as if they assumed that their role was to feed them and then step back out of the way while they grew. The little ones spent a lot of time sleeping, which I assume is typical of newborns of all species.

May 4: The little ravens are growing so rapidly that they seem to be changing as we watch them! Each day they are getting bigger and darker—and stronger. This morning there were more squawks from the nest. One of them is definitely going to be a talker—taking after its mother.

This evening with binoculars I looked more closely into the nest when the adults were feeding the little ones, and I discovered that there are actually six little nestlings! I couldn't believe it. How could I have missed seeing two additional beaks all this time? The little ones are all now big enough that when everyone put its little head up, six little beaks can be easily seen. What a surprise! A few days ago, I thought I had detected movement of a fifth little one, but I was sufficiently unsure that I didn't note it.

May 5: There is not much new to report today in Ravenworld. It appears that during this stage of development the adults are simply going about their chores of feeding the little ones and waiting for them to grow up so that they can then begin the next phase of training. There is a lot left to do, but it will have to wait until the little guys are ready to fly. The adults, meanwhile, have been very attentive—tending to the regular feeding cycles, cleaning up the nest, and driving off the hawks. With six little beaks to feed, their world is very busy.

May 6: The nest seems to be getting darker by the hour as the little ones' feathers continue to grow out, and its limited space getting much fuller as the little ones themselves grow. When the adults feed the nestlings, they stand on the rim of the nest, as there does not appear to be much room elsewhere. Little beaks seem to poke up everywhere, crying, "Feed me!" Everyone seems to get fed, however, for when dinnertime is over, everyone goes to sleep. This has been fun to watch. In three or four weeks or so, they will probably be getting ready for their first flight-training lesson!

May 7: This morning when I took some kibbles and chicken pieces out to the ravens, both were sitting near the top of the larger eucalyptus tree. I called to them as I usually do, and before I had finished putting down the food by the pine tree, Freya flew over. I looked up and acknowledged her, and as I called again, she hopped to a closer branch. I stepped back a couple of feet so as not to intimidate her and called again. She flew to another branch closer to where I was standing. Again I stepped back and continued to talk to her. This time she

We're still hungry!

flew down to the ground, where I had spread out the food. She approached it cautiously while I was still standing there. I stepped back another few feet to ease her trepidation, and she started pecking at the morning's offering while keeping an eye on me. This was the first time that she had voluntarily approached that closely. It appeared that she is coming to trust me even more!

As for the little guys in the nest, they are getting bigger and stronger. This morning while one of the adults was perched on the nest rim, two of the little ones stood up and flapped their wings. Their feathers are getting longer and darker. When they are all together and asleep, they now fairly well fill the nest. Both adults spent considerable time tending to them this morning. I was able to get a couple of photographs of the little ones after the adults left. One photo shows two in the foreground with their little beaks open.

May 8: This morning I had a very similar experience with Freya to what I had yesterday morning. As I put out the chicken pieces and kibbles while calling, she flew over while I was still there and stayed while I talked to her. Again I stepped back to appear less threatening, and this time she hopped across the branches, flew down to the ground directly, and started eating. I guess with six mouths to feed, one gets braver!

The older little ones are now occasionally standing in the nest and stretching while flapping their little wings. They are starting to look like miniature adults. They still open their orange beaks for their parents, but they are growing fast. When they are all bundled together asleep, the nest becomes a dark pool. Over the past few weeks, their collective color has evolved from light gray to nearly black.

This evening there were a total of five people visiting to have a look at the little ones, and as one might expect, all this attention made the adult ravens quite nervous. While we were all standing on the balcony with our binoculars and cameras, the adult ravens stayed away from the nest. We noticed this reaction and decided to make our presence less threatening to the ravens by going inside. When we retreated to the living room, where we could still see the nest from the window, everything returned to normal, and the adults continued with their feedings. Everyone took several pictures.

May 9: The little ones are growing so rapidly that one can almost blink while looking at them and notice a change. The "older" ones are now standing up more frequently, stretching and flapping their wings. They all pile together with their little beaks poked out ready to "receive" in the off chance that one of the adults—or anyone else for that matter—happens by. But they spend the majority of their day asleep. Their job now, so it seems, is growing and sleeping.

May 10: More of the little ones are now occasionally standing up in the nest and looking around at the big world out there. The tops of their little beaks are also getting darker. The insides of their mouths, however, are still orange with yellow rims. The adults have no problem seeing that target when it is open! Meanwhile, the nest is getting blacker and fuller with each passing day. And there are fewer bare spots exposed on their little bodies, as their pinfeathers are developing into real feathers.

The adult ravens, in the meantime, are taking all of this rather nonchalantly. On breezy days, they go soaring in the wind, playing in the sky. They spend much of their time perched near the top of the large eucalyptus, preening. But when the time comes to feed the little ones, they are quite dedicated. They make numerous trips until all the little orange beaks are closed. And they know where every food source is. Not only do they have our reserve of kibbles and chicken, but our next-door neighbors are also getting caught up in the excitement of hatchlings and had been leaving out hamburger pieces for them.

Ravens will take all of the food that's available from a given source, and anything they don't eat immediately they will carry off to a storage cache. These intelligent birds plan for the future by caching their surplus food in a secluded place known only to them for later access. They normally maintain a wide variety of caches in multiple places, and they remember the location of each. Ravens have been observed pulling up segments of sod, inserting their surplus into the hole, and then replacing the sod. They will poke food into spaces they dig or into crevices they find, and cover their stash with leaves and twigs and any other available material.

Our ravens have at least two cache sites that I know of. One is up the canyon to the east, and the other is near the western end of their territory. And undoubtedly they have several others.

May 11: At least two of the little ones were squawking first thing this morning as I went outside to leave some kibbles and chicken for the adults. These little guys are growing so fast that were I not keeping a daily log, I would not recognize their changes. Their feathers are developing nicely and growing along with them. It appears that one of the nestlings may have a row of white feathers, as did one of the juveniles of last year. But this has yet to be verified, as most of the time the nest is a jumble of black feathers where the little ones are all piled together asleep.

It's crowded in here!

May 12: There was more squawking from the nest at feeding time this morning. I was a little later getting out there today, and Freya flew over to the pine tree as soon as I called. She apparently was hungry. She approached me closer today than she'd ever done before.

The little ones are growing at an astounding rate. This afternoon one stood up in the nest and flapped its wings for nearly thirty seconds. It has grown to at least half of adult size. They have made incredible progress.

May 13: If one could say that ravens go through a "teenage" stage, then our little ones in the nest are beginning to look like teenagers. Now three or four of them are trying to stand up and stretch their wings, all pushing and shoving with a flurry of feathers. There is so much flapping and squirming going on in the nest that I wonder if there will ultimately be enough room for everyone!

May 14: More and more of the little ones were standing up in the nest flapping their wings today. One of them still appeared to have a row of white feathers along the trailing edge of it wings, but I have not been able to verify that. The nest is looking fuller with each passing day. It has now become almost a sea of black—except when the little beaks are open. The original orange color is now becoming more pinkish. But it is still an obvious color for the adults to see. And again, at this point their job is to eat, sleep, and grow.

May 15: This evening, there was a predator threat from the ground. Because we live adjacent to a national forest, we have every species of wildlife found in Southern California—and in this case, specifically bobcats. Although we have seen a bobcat off and on several times over the years, the fire of last August destroyed so much habitat in the mountains that they have extended their foraging range. As a result of the fire last year driving out the predators, we have had an overabundance of ground squirrels. This past spring the bobcats discovered this. For quite some time now we have been routinely spotting bobcats on at least a weekly basis. When the ravens catch sight of a bobcat, they dive-bomb it, missing it by inches, and when they do this, they call out with a strong scolding call that easily translates as "Bobcat! Get out of my territory!"

Looking for lunch

This evening the ravens were diving at something at the foot of our hillside and scolding while they dove. I suspected that it was a bobcat they were attacking and went out onto the balcony to investigate. Sure enough, I saw it walk under one of the oak trees to protect itself from all the harassment. I watched this action for a while and then went back inside. But the ravens' scolding didn't stop. About ten minutes later, I went outside again to see what was keeping their interest, and I was surprised to see that there now were two bobcats! One was lying down on the grassy knoll, ignoring the ravens' assaults, while the other was looking toward the hillside for ground squirrels—and also trying to ignore the ravens. The ravens' bombardment of the poor bobcats was relentless—and it also drove away all the squirrels. When the bobcats finally realized that dinner would be delayed today, they slowly walked away. I had a camera handy and took many pictures of these proceedings.

May 16: The nest is now starting to look like it is being overburdened with way too many occupants! The "teenagers" are filling it more and more each day. They take turns bumping into one another and standing up and beating their wings. Their wing feathers have grown to the extent that they seem nearly full-size. One of them likes to stand on the edge of the nest and beat its wings, and once it nearly lost its balance. Our teenagers are not quite flight ready, and it would have been a long drop to the bottom of the eucalyptus.

May 17: The young ravens are getting ever more restless in the nest. They are doing a lot of shuffling, wing flapping, and preening, as if they are getting anxious to get out into the world! I suspect it is taking more to feed them as well, as the parents are making several trips per day bringing food to the nest.

May 18: One of the bobcats came back today, and the ravens immediately spotted it. The bobcat was coming down the hillside by the mailbox in front of our house looking for a hapless squirrel when the ravens noticed it and began their dive-bombing attack—again missing by mere inches. The bobcat walked away to avoid the harassment but returned shortly thereafter—and was spotted again. This time it continued down to the garden while staying under the bushes to avoid the ravens.

In the anthology of ravens, the usual relationship between ravens and predators is to team up. The classical cases are the partnerships between ravens and wolves, where there seems to be an uncanny social bond—so much so that the raven is often called the "wolf-bird." Where the wolves go, ravens regularly follow to scavenge their kills, immediately feeding on the scraps. Unintimidated by the wolves, they will often feed alongside them. And the ravens return the favor, especially in cold climates. Ravens may spot a frozen carcass of a dead animal in the woods. Since they don't have the strength to tear it apart, they will seek out a pack of wolves and lead the wolves to it. The wolves tear into the carcass and generate pieces small enough for the ravens to eat. At the kill site, the ravens also serve as additional eyes and ears to alert of pending danger.

The relationship between the two species appears to be both social and symbiotic, where each has reached a state where one is rewarded by the other, with each seeming to be aware of the other's capabilities. A possible basis of this relationship, as suggested by Bernd Heinrich, is that ravens lead wolves to prey and alert them to dangers, and are hence allowed by the wolves to share in the spoils as a reward.[20] Since it is known that ravens can be attracted to wolf howls, it may be possible that wolves also recognize certain raven calls that indicate the presence of prey.

---

20 B. Heinrich, *Mind of the Raven* (New York: Harper Collins, 1999), 232–44.

Ravens have been known to follow other predators also, including humans on deer or moose hunts. So our ravens attacking predators seems somewhat out of character. Freya and Tyr may have had an unfortunate experience with bobcats, especially since bobcats have an ability to climb trees. It's very possible that our ravens know this and were doing everything in their power to discourage the bobcat from going after their little ones.

May 19: The "little ones" are now essentially young adults. They are standing up and moving around the nest much more frequently now. They are taking turns standing on the edge of the nest and flap-

How do these things work?

ping their wings—even to the point of occasionally nearly losing their balance and falling out. They appear ready to go at any time! It will be most interesting to see how the parents judge their flight readiness and start their flight training.

This evening I took a video of the teenagers standing on the edge of the nest and flapping their wings. All of them seemed to be curious about looking over the edge, as if to say, "It's not really all that bad!" They will definitely be leaving the nest soon.

May 20: The little ravens are really raring to go! They are now spending much more time standing up and taking turns climbing out onto the edge of the nest and flapping their wings. They are ready to take on the world! One of these days in the very near future, they are going to find themselves airborne. For all practical purposes, they now look like adults—with the only giveaway being their red mouths and the lighter coloration around the outside of their beaks.

May 21: There is nothing much different to report from the nest today other than that the little ones are getting ever more restless. Any day now they will be coaxed out of it by their parents. I have no idea what cues the parents that the little ones are ready, but clearly something does. We are going to be out of town the next two days, and it's very possible that one of the older ones will take its first "leap of faith" during that time. Although I hope not, as I really would like to be present for this big event. One of them appears nearly ready, however, as it keeps hanging precariously close to the edge of the nest and flapping its wings as if daring gravity to pull it off.

May 25: The adult ravens were very glad to see me with the kibbles and chicken this morning after our three-day absence! There was no new flight activity from the nest today, although the juveniles seemed to have spent an inordinate amount of time standing and flapping their wings. They appeared to have been taking turns standing on the edge of the nest and stretching. Each one would extend one wing at a time, stretch it down over the side of the nest, and hold it out for a few seconds. Then it would switch to the other wing and repeat the process.

CHAPTER 8

So Much to Learn!

• • •

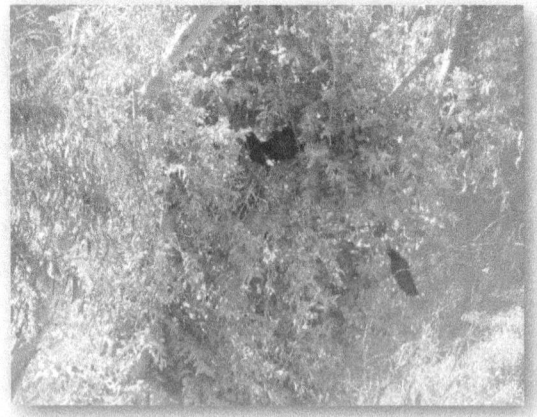

The first "leap of faith"

MAY 26: THE FLIGHT TRAINING has begun! This morning two of the raven juveniles were out of the nest. One of them was on a branch to the south of the nest, and the other on an easterly branch somewhat behind the nest tree. Neither was very far away from home. While I was watching them, the one to the south "flew"—or more accurately, glided—over to an adjacent oak tree. One of the adult ravens followed it while talking to it. As I continued to watch, I counted the juveniles remaining in the nest and confirmed that there were indeed four left. This was the first time I was able to verify with certainty that there are beyond a doubt six juveniles. Since the first time I thought I saw six beaks, I have not been able to rigorously confirm it. It is very difficult to isolate individuals from the squirming black mass of feathers.

I looked out again about a half hour later, and one of the juveniles was on the ground over by the pine already. An adult flew down to it and then coaxed it to fly across the dry creek to one of the oaks on the far hillside—which it did.

This evening two of the juveniles, likely the same two as I saw this morning, flew over to the top of the berm near our balcony. One of the adults was coaxing them. They stayed on the ground for a few minutes and then both took off. One flew directly to the nest, while the other followed the adult and flew to the hillside across the dry creek.

It's obvious that the eggs were laid a few days apart and hatched on a similar schedule, and this timing difference will spread out the maturation period for each juvenile to leave the nest. So sometime over the next few days, these first two should be joined by one or more of their siblings. It has now been nearly six weeks since I had suspected that the first eggs were hatched. So it seems that the little ones are essentially ready to fly six weeks after coming out of the shell!

The fox came out this evening to eat some of the kibbles I had left for the ravens. I stepped out onto the balcony to take a picture of it. As I stepped out, Freya flew over to the pine tree, landed on a lower branch just above the fox, and then called as if she were talking to me. It was her very nonthreatened "Over here!" call. I took a couple of pictures of the fox and then walked to the edge of the balcony. She stayed on the branch and called again while looking at me and then the fox. I got the fox's attention and encouraged it to leave. As it walked away, Freya seemed satisfied and left.

May 27: This morning, a fourth little raven left the nest. It was under a branch about three feet below the nest, testing its wings for a short flight to the nest tree. I watched while it made its first "leap of faith" as it left its home tree for the first time. I had somehow missed the third juvenile's test flight and am sorry that I did. The two remaining nestlings were pacing nervously and testing their wings.

The day was overcast and drizzly for the majority of the day, so the new fledglings were difficult to keep track of, as they tended to want to tuck under branches to stay dry. They did very little actual flying, since they had insufficient strength development. The parents were very attentive, watching out for where everyone was.

This evening marked the grand finale! The remaining two little ravens left the nest! They were perched on a lower branch, getting up their courage to make their own first leaps of faith away from home.

So now begins the next phase: where the adults train the juveniles on how to survive on their own. They have much flight training ahead of them, and they must learn how to find food and how to survive in the cruel world, where they too can find themselves as prey. It occurred to me a few days ago that I have not heard Freya call out with her full enthusiastic warble for several weeks—the warble that says, "I'm queen of the mountain, and all is right with the world." With six kids to train and feed, she is probably staying focused on the task at hand—celebration will have to wait!

May 28–June 1: We were out of town for a trip to Albuquerque during these five days and missed the very early flight-training stages of the juvenile ravens. However, our next-door neighbors, Bob and Darlene, kept track of their progress, and this is what they had to report:

May 28: One of the youngest ravens returned to the nest. The other juveniles, meanwhile, continued their training.

May 29: The nest was empty, flight training was underway, and things looked good for the raven family!

May 30: Another day dedicated to training, and all was quiet.

May 31: Around six in the morning, an adult raven perched on the neighbors' wooden railing squawking loudly for food. The neighbor put out some beef pieces, and the food was gone after each raven made several trips. Around eleven o'clock, there was a major commotion, where the adult ravens were very upset. Our neighbors checked for the source and found a bobcat below their deck area. They made a loud noise, and the bobcat ran east into the vegetation. All was peaceful again.

June 1: The bobcat came back again around two in the afternoon, but this time it was on the east bank of our hill-slope opposite their yard. Our neighbors heard the ravens squawking and went over to investigate. They said the ravens were diving at the bobcat and squawking like crazy. The bobcat headed down the hill and into the bushes.

June 2: I am now back from our short trip and returning to log-book duty. It looked like the adults had all the "cadets" stationed in the easterly section of the canyon for their basic training today. The adults were carrying food in that direction, and they seemed to go to the oaks that are near where last year's nest was. I have suspected for a long time that this location is where the ravens roost for the night. So maybe learning about this region of their new world is also part of the early training for the cadets.

It will likely be a couple of weeks yet before I see them all flying together. The little ones are still working on developing their muscles, so all of their flying so far has only been for short distances. But I do hear the adults calling cadence during the flight lessons. Flight-training segments at this stage are fairly short, with many rest breaks.

I mentioned a few days ago when I heard Freya warble that she hadn't done this for quite some time. Well, this morning she flew over to the pine tree and again warbled. I heard her, stepped out onto the balcony, and called back. She warbled once more! We talked back and forth for about five minutes. She seemed happy, so apparently all was going well.

June 3: The juveniles remain relegated to basic training up the canyon, although this morning one of them came down and perched upon the eucalyptus tree near its nest while the adults picked up the chicken scraps I left out. Again, flying lessons were short, with many rest breaks. At this stage of the youngsters' development, these lessons seem to be conducted more as a one-on-one procedure rather than as a group formation. As an example, one of the adults was leading a juvenile in flight along the far side of the canyon, calling to it all the way. The other adult, meanwhile, was soaring overhead. But soon the young cadets will have enough strength for longer ranges.

June 4: Today there was very little trace of the juveniles, as they were still stationed up the canyon undergoing "basic training." I heard an occasional squawk that was definitely from a young raven, and once in a while I would see one flying a short distance. But since the cadets are all still in basic training, it will be a while yet before they come out in squadron strength. The adults, in the meantime, were very active in finding food and made several passes over their regular spots. They are still feeding the juveniles, even though the youngsters are now formally out of the nest.

June 5: There was little sign of the juveniles again today, although I heard several young raven squawks across the dry creek. It is very difficult to observe their ongoing activities at their present location. But the parents clearly know what they were doing for this stage of the required training and development. The adults continue to be busy making food runs for the youngsters.

June 6: This afternoon two of the juveniles were perched in the eucalyptus tree near their nest. They squawked at each other and did short flights to some of the other branches. At approximately the same time, one of the adults was giving another young cadet turn-and-bank flying lessons. It was fascinating to watch as they flew in a zigzag pattern.

June 7: This morning the adults had all the "trainees" on the ground in the dry wash at the bottom of the canyon. I'd seen this element of training in previous years. It appeared that the youngsters were learning to pick up pebbles for their gizzards and/or were finding small food items like insects and grubs. It's also very likely that the adults were teaching the juveniles how to cache food by practicing simulations with inedible objects. The adults were talking to the juveniles in a modulated throaty call, as if they were giving instructions. This call, again, was unlike any of the usual raven calls we normally heard.

The little ones are flying more easily now, but at the same time they are all just a little hesitant about going very far. They are still working on building their muscles, and they are making very good progress.

June 8: I didn't see or hear much from the little ones today, other than a few squawks from across the dry wash. The adults, however, are nevertheless patrolling their regular food sources, as they still have the responsibility of feeding the juveniles.

June 9: Today was a breakthrough day for some of the young cadets. Late this afternoon, the adults had at least four of them out flying over the eastern section of the canyon, and they brought one of them—likely the eldest—to their favorite pine tree, where all three circled and then flew back toward the eastern canyon ridgeline. The juvenile, meanwhile, perched on a branch of a nearby burned-out oak, while the adults continued to fly across the canyon to pick up the other three youngsters that had been "parked" temporarily. Both adults then flew with the other three for several minutes, soaring up and down the canyon wall until the young fledglings got tired and had to stop. I was not able to verify if the remaining two youngest juveniles were part of this exercise. It is very likely that they are still a few days behind this specific maturation point, and I probably will not see them joining in with this new flying regimen for few more days.

Later in the evening, after I had already left the ravens some kibbles and chicken pieces, I heard a different call from the pine tree and an unfamiliar response. I stepped out onto the balcony to investigate. I think what was occurring was that an adult raven was in the pine and was calling to the juvenile to come over to where the food was. The adults need to teach the juveniles how to find food, and in past years they had always brought the juveniles over to the food source to teach them to feed themselves. So I should soon be seeing the youngsters joining the adults under the pine tree.

June 10: Late this afternoon while I was out on the balcony, I noticed a raven perched across the canyon on the branch of a burned-out oak and facing our way. I suspected it might have been Freya, as she will occasionally perch there, where she can survey all her usual food sources. From that vantage point she can see the locations where I normally leave kibbles and chicken, and where our neighbors also put out raven-tempting morsels on their railing. I waved and called out to her. And sure enough, she came off her perch and slowly soared toward me. She swooped fairly low over the pine tree, while checking to see if anything was there, and then drifted overhead and over the house. As she flew over our house, she warbled to me as if to say hello. She continued over to the neighbors' house and came to rest on their wrought-iron railing. She then hopped over to the wood rail, where she disappeared from my sight. Very possibly there was some food there.

When I spread out the evening kibbles and chicken, I was treated to a fascinating sight. Once again, in the same manner as last night, the adults brought one of the juveniles over to the pine tree, where it slowly circled and then flew back across the dry creek toward the canyon and perched on the same burned-out oak on the hillside. One of the adults then picked up some pieces of chicken and flew over to it—and fed it! The juvenile spread its wings and flapped as the adult approached, and opened its mouth as if it were still in the nest. That was its body language telling its parent that it was hungry. It is not yet able to feed itself, but that day will be here very soon!

June 11: This morning there were three ravens on the ground at the pine tree! The adults brought one of the juveniles over to show it how to forage, likely the one that flew over last night. This was the first time that one of this year's youngsters had actually landed on the ground where the food was. I suspect that this particular juvenile was the eldest, as it seems to be at a more advanced stage of its training.

This evening four of the little ones were flitting around the eucalyptus tree near their nest. On the whole, today all six have more or less remained in that general vicinity or on the hillside across the dry creek. I could hear their squawks from that location. Their little squawks have become more and more frequent with each passing day.

When the adults came over for their chicken pieces, one of the juveniles flew partway with them but didn't land there. Instead, it turned and flew south to the burned-out dead tree north of the flood-control catch basin.

Mom and Dad supervising Junior

It perched in that tree and then started to explore the branches. It picked up a twig about six inches long and started poking it at different parts of the limb it was on. It did this for a couple of minutes. Apparently seeing one of the adults approach, it dropped the stick, spread its wings, and squawked while flapping with short, rapid strokes. It was hungry! As the adult landed beside it, the little one opened its mouth, and the adult put some food into it. With that accomplished, the adult flew back to the pine tree for more food, while "junior" stayed where it was.

I could hear similar activity going on out of sight on the far side of the eucalyptus trees, as it was obvious the adults were feeding the other juveniles. With six beaks to feed, I do wonder how the parents remember to feed everyone! And how do they find them all?

June 12: Today's raven story is quite exciting. I was out on the balcony checking into the source of all the squawks from the juveniles when one of the adult ravens came flying down from the canyon. From its behavior, I suspected it was Freya, which turned out to be the case. She slowed down to nearly a hover as she approached. I waved and called to her, and she turned toward me and very slowly flew by me at a distance of a few feet. I waved and called again as she went by, and she warbled as if to say hello. She flew south for a short distance and then turned and flew back—and repeated the same act. She again flew very slowly and very close—and when I waved, she warbled. She then went over to the pine tree and hopped down to the ground where there remained a few kibbles.

Meanwhile, the juveniles are still hanging out on the other side of the dry creek behind the eucalyptus trees. The adults seem to be spending more time with them. They will go over to the cadets and "communicate" with them with a different, throatier call. Likely the parents are instructing the youngsters in life's basic lessons—how to avoid predators, how to find food, and what size stones to use for one's gizzards. The juveniles haven't traveled very far yet, as their muscles are not fully strengthened. But I do occasionally see them fluttering across the branches of the trees at the base of the dry wash. And one of them seems to like the burned-out dead tree in the south by the flood-control basin. Any day now I should see them all engaged in longer flights.

When the adults went over to the pine tree to pick up food this evening, they called for the juveniles to join them. Again, it was a different kind of a call—more of a "Come here" command. When no one showed up, they each hopped down to the path, picked up a beakful of chicken scraps, and flew over to the juveniles. I could not see the juveniles, as they were in the trees beyond the eucalyptus, but I could hear their enthusiastic squawks as the adults approached with dinner!

June 13: Today the adults took at least two of the juniors for a flight up along the ridgeline of the far canyon. I watched them with binoculars for quite some time as they gracefully soared, with the youngsters occasionally stopping to rest. This evening the adults had at least five of the juveniles—and probably all six, as I undoubtedly could have missed one—flying along the flood-control catch basin. Because the flood-control basin is a fair distance from where I was standing on the balcony, I was only able to see five cadets as they occasionally rested during their training. But the last little guy had to be there somewhere.

June 15: I was awakened this morning shortly after dawn and long before I was ready to get up by a juvenile squawking from the pine tree outside. The adults were introducing the youngsters to one of their known food sources and had been bringing the eldest one around fairly regularly. However, one of his lessons in life is going to be that if he insists on getting up a lot earlier than I do, then in the mornings he is going to have to be patient. By the time I actually did get out there to give them their breakfast (a couple of hours later!), all of the ravens had disappeared. No one was in sight. I called but didn't hear any return calls. I called again and started to walk away. But true to form, both adults came flying in from somewhere, and Freya flew directly to the pine tree.

June 16: This morning when I went out to give the ravens their morning ration, there was no one around. I called to them and then started to cut up the chicken. When I finished, I looked up, and there was Freya, sitting in the pine tree waiting for me. As I approached with the morning meal, she never moved. Instead, she waited for me to spread it out and take a few steps back. At that time, she hopped down to the ground and started collecting for the little ones.

The ravens seemed to be unusually hungry today. Around midafternoon, one of the juveniles was perched high in the large eucalyptus and squawking. It would squawk and rapidly flap its wings, its body language saying to its parents: "I'm hungry!" I immediately went out and gave them some more kibbles. When I did that, I waved to the youngster. While I was waving, I had not noticed Freya fly over to where I was standing. When she got my attention, I stepped back a couple of paces, and she immediately picked up some kibbles and took them to the pleading juvenile. It was most grateful!

Later I cut up some chicken for the evening meal and took it, along with some more kibbles, over to the base of the pine tree. I called to the ravens, and Freya immediately flew over to where I was. Often in the evening she will approach only as far as the eucalyptus tree, where she will patiently watch and wait. But today she was hungry. Maybe there was a scarcity of naturally occurring food in the hills today. She and Tyr very quickly cleaned up all the chicken and most of the kibbles. The squirrels ate the remainder.

Meanwhile, I watched four of the juveniles practicing aerobatics around some of the burned trees east of the flood-control catch basin. They would fly and dive, tuck and turn, and then rest. This continued for a half hour or more. They were having fun!

June 17: At feeding time this morning, Freya once again flew over to where I was standing when I called. She would approach to a distance of about ten feet but was cautious about coming much nearer.

Later in the afternoon, I did finally see all six juveniles flying together. They were doing aerobatics over the flood-control basin, while the adults were looking for food around both houses—ours and the neighbors'.

Kibbles and chicken bits

June 18: This morning I was awakened around six o'clock by squawks outside the window. I got up and investigated, and it seemed that the adults had again brought one of the juveniles over to where we put kibbles out for them. All three were on the ground pecking, and the juvenile must have been wondering what was going on, as it kept squawking. I suspect that this might have been one of the younger juveniles, who has reached this point in its maturation where it is time to learn more of the intricacies of foraging. However, I am not in the habit of giving them food quite that early, so they were ahead of schedule!

By midmorning, four of the juveniles were practicing formation flying and aerobatics. They flew two by two in tight formation for a while, and then they would group together and tuck and dive. I watched them for probably twenty minutes until they finally grew too tired and had to stop to rest. They were really having fun!

All throughout the afternoon, more of the juveniles were flying around overhead—and squawking. It appeared that they are getting stronger and able to sustain longer flights. Several landed in the eucalyptus tree where their parents like to hang out.

June 19: This morning when I was getting ready to take out some chicken and kibbles, none of the ravens seemed to be around. I looked for them in their typical perches and called to them, but I didn't get a response as I usually do. I didn't see anyone. But as I walked over to the pine tree with the food, behold! There in front of me, perched on the lowest limb, was Freya silently watching me! She just sat there while I took everything over and spread it out on the path. I never saw her fly in. She let me approach to well within ten feet of her and never moved. As I backed away a respectable distance, she hopped down to see what the morning offering was.

June 20: Today the cadets were in more extensive flight training. The youngsters are inclined to prefer the burned-out, dead tree north of the flood-control catch basin and often perch there. Because they tend to migrate there fairly often, we have informally called this tree the "training tree." This tree stands alone, devoid of any obstructions, and gives them a good view of the canyon entrance. This afternoon one of the adults and four of the "trainees" were perched there when the adult decided to leave. And just like a precision squadron takeoff, one after the other left the branch in trail. That was an imposing sight!

This evening, after I had cut up some chicken and taken it and some kibbles to the usual location, I called for the ravens but got no response. They were nowhere around. I went back into the house, and after a few minutes I heard them return. I went outside and waved and called to them, and the whole family flew low and slow overhead with the juveniles squawking all the time. All eight were flying in formation. It was a spectacular presentation. They glided around for a short time, and then all of the juveniles flew down to the training tree and perched. I got the camera and took several pictures of all of them together. In the meantime, both parents had flown over to the pine tree where I had left their evening buffet, and a most fascinating event then took place.

The parents perched on the lower branches of the pine, knowing that there was food below them. But they didn't immediately go down to pick it up. Instead, they wanted the juveniles to come over and learn more about how to collect it. Freya started calling to the youngsters with a "Come over here!" call, which is similar to the "Where are you?" call but at a quicker cadence. She called for several minutes, but nobody from the training tree budged. She called again and again, but still nobody came. After a while, her call got a little more agitated, as if she were saying, "If you don't get your butts over here, I'm going to get out the paddle!" It would seem that youngsters of all species are similarly behaved! Finally, one by one, the juveniles started to fly over to their parents, some landing in the pine and some in the eucalyptus. Then, and only then, did the parents hop down to the ground to pick up food. It was as if they wanted to show the juveniles how to do that. When the youngsters saw them picking up food, they squawked and rapidly fluttered their wings to say, "I'm hungry!"

June 21: This morning when I went outside, the whole "squadron" was up and flying high over the canyon. All eight ravens—the two adults and six juveniles—were soaring over the ridgeline! It was an impressive sight! As I put the food out and called them, they all migrated over to where I was. The adults came to the pine tree, and all six of the juveniles landed in the tall eucalyptus. And they were noisy!

The youngsters have now developed the strength for longer flights, and everyone had gone traveling for most of the day. They came back temporarily in the evening as I was spreading out some chicken scraps and kibbles, but then they disappeared again. I suspect that the adults were taking the cadets for a longer, cross-country flight. Around seven in the evening, they all came back from "somewhere."

June 22: Today the whole family was generally around. Many of the cadets were practicing flight maneuvers in groups of two to five—catching the wind, drifting, and diving. They are definitely getting stronger and much more skillful. The adults, in the meantime, are busy trying to keep them fed. Six beaks require a lot of food! Freya is getting braver in letting me approach her more closely. She knows that I am a source of food and not a threat. Tyr is still a bit more cautious, but he is nowhere nearly as skittish as he used to be.

This evening while I was updating this log input, I heard Freya outside suddenly start her "Where are you?" call. It was about the time of night when I normally give them some chicken and kibbles, so I went out on the balcony to see where she was. She was in the top of the eucalyptus tree facing the house and calling straight at it. It appeared to be quite clear what she wanted! I went outside to cut up some chicken, and she watched me from her perch.

But partway through the preparation process, Tyr spotted the bobcat on the far hill and started diving at it all while yelling his "Bobcat, get out of my territory!" call. This call is quite different from telling another raven to get out of his territory, and it uniquely refers to a bobcat. Freya heard the commotion and flew over to join in. When I finished with the chicken and had everything together, I called to the ravens, took their evening meal over to the base of the pine tree, and spread it out on the ground. I continued to call but got no apparent response. I was about to give up, thinking that they were probably more interested in where the bobcat had gone, when I looked over to the pine tree. There she was, perched on the lowest limb, looking at me and wondering what all the fuss was about. She apparently had flown in from the ravine while I was scanning the sky.

June 23: There is not much new to report from Ravenworld today. The juveniles are definitely getting stronger and thus are flying around much more often and taking longer flights before resting. Many of them are now flying between the eucalyptus trees and the pine where they can find food. The eldest juveniles have learned what this process is all about. All six like to hang out in their training tree and spent considerable time there today.

June 24: The parents are trying to teach the young ravens how to scratch for food themselves. Both adults had two of the juveniles on the ground by the pine tree early this morning, teaching them how to identify edible items. The juveniles were hopping around, scratching and pecking.

The juveniles are definitely flying more and are now more inclined to land in the eucalyptus trees where the adults go. Freya's favorite perch is near the top of the largest eucalyptus, and she is now routinely being joined by one or more of the juniors.

June 25: I photographed all six juveniles perched in the training tree this morning. They were all together on the same side of the tree.

The adults are getting more serious about showing the young ravens how to find food on their own. Once again, as with this past Sunday, when I put chicken bits and kibbles out for the

Cadets in the training tree

evening, Freya flew over to the lower branch of the pine tree and started calling to the youngsters to come over. Tyr stayed in the eucalyptus, watching the proceedings. At no time did either raven go down to the food while Freya was calling. She continued to call for what seemed like ten minutes. Meanwhile, we could hear the juveniles squawking across the canyon. Finally, the youngsters started to arrive at the pine tree where their mother was perched. Then and only then did she hop down for the chicken, and she was immediately followed by two of the juveniles. In the meantime, several ground squirrels had eaten much of the kibbles that were laid out. Freya chased the squirrels away to make room for her trainees. The little ones stood there and squawked while Freya picked up some pieces of chicken. One of the juveniles then spread its wings and beat them with rapid short beats and squawked, telling its mother "I'm hungry!" She went over to it and put some food into its mouth.

Although the adults are still feeding the youngsters, they are also trying to show the juveniles how to forage on their own. This is one of life's skills that all ravens must learn in order to survive. The method used by the parents to teach the youngsters this critical skill is most interesting. Apparently, once the adult ravens have identified a known food source, they call the youngsters over so that they too will learn to recognize it. Then they show the youngsters the process of eating it.

June 26: The whole gang was over at the feeding area this morning—flying between the pine tree and the eucalyptus. It was quite a sight with all of the ravens flying very slowly and low overheard—and squawking all the time. The little ones as yet do not have the subtleties of the raven language down, but the parents apparently are able to understand them.

A couple of the juveniles joined the adults on the ground to peck at the chicken morsels and kibbles. The remaining youngsters stayed behind and squawked while rapidly beating their wings, telling their parents they were hungry.

Around noon I heard Freya calling and many of the juveniles squawking. I stepped out onto the balcony to investigate and concluded that they were probably hungry. I took some kibbles over to the path by the pine tree, spread them out, and called to the ravens. Two immediately flew over to the pine— Freya and one of the juveniles. As I went back into the house, what looked like the rest of the family arrived. Freya and two of the youngsters hopped down to the kibbles. Meanwhile, several ground squirrels started gathering around also, as they too like kibbles. Freya chased the squirrels away so as to give her kids a chance to eat. The youngsters started pecking at the kibbles and were joined by two more juveniles.

They are learning to forage on their own. Possibly within a few days the entire family will be on the ground looking for food without need for strong encouragement from the adults. The parents will likely still be "hand-feeding" the youngsters for some time yet, but it does appear that they are being weaned. In the meantime, an adult and one of the younger juveniles flew over to the broken limb of the eucalyptus tree, and the youngster beat its wings and squawked to the adult, saying, "Feed me!" The adult then fed it.

June 27: This morning all six of the youngsters were perched in the pine tree. The majority of the juveniles have by now learned how to get food for themselves, although they are still looking to the adults for guidance. Some of them still rapidly beat their wings and squawk when they see an adult that might be ready to offer a meal.

I arrived home this evening about an hour later than the time I normally leave out the chicken and kibbles for the ravens. All six of the juveniles were squawking and hovering around the pine tree, waiting. It appears that they have definitely learned where at least one food source is!

June 28: This morning around seven o'clock, we were treated to a cacophony of squawks from the juveniles as they flew from the pine tree to the eucalyptus, wondering where breakfast was. It is obvious that their lessons on how to find food have indeed been going well!

Today I noticed one of the ravens—I couldn't tell which one, or whether it was an adult or a juvenile—return to the nest twice and behave as if it were repairing it.

The juveniles were quite active in the sky today. There was a slight breeze, and everyone was taking advantage of it. The whole family was airborne and catching the updrafts. There were a lot of ravens in the sky!

Two or three of the little ones apparently have inherited their mother's tendency to talk a lot, as they are quite vocal. Now that they are getting older and stronger, they seem to be squawking all the time. But so far, all vocalizations are just short squawks, as the juveniles have not yet learned the adult raven "language" of different calls.

June 29: All six of the juveniles were on the ground by the pine tree this morning, learning to pick up food. Most were doing a good job, but the youngest among them were still begging their parents by rapidly beating their wings and squawking. The parents would relent and feed them. This lesson will continue for at least a few more days.

For a significant part of the morning, all the ravens were somewhat agitated. There may have been another lesson going on, as the adults were calling to the juveniles, and the juveniles were darting around. Earlier, from the calls the adults were making, it sounded like the bobcat was in the area and they were trying to drive it off. Maybe they were taking advantage of this event to teach the youngsters. Ravens are such complex birds that there is a lot a young fledgling must learn before it ventures off on a life of its own.

Meanwhile, the juveniles have now grown so close in size to the adults that it is very difficult to tell them apart. Each youngster's mouth is still red inside, however, and the outside edges of their beaks are still lighter than that of an adult. So if one can see their faces when they fly over, then one can readily tell who the juveniles are. However, one must be quick when observing, as there are a lot of birds present when they are all up and flying around!

June 30: A few of the juveniles have learned to gather food on their own. Some were picking up kibbles and pieces of chicken and flying off, while one or two were still begging from their parents. They would stand on the ground near the food while rapidly beating their wings and squawking until a parent fed them. Although I could not confirm how many were doing this, as of this morning I only saw two, and I strongly suspect that these were the youngest of the six. They are still a few days behind their older siblings in the "Lesson Plan of Survival."

Our next-door neighbor witnessed a most amazing phe-nomenon around one o'clock this afternoon. She walked out onto her deck and saw both adults and all six juveniles fly in from the west and toward the southern end of the eastern ridge by the flood-control catch basin. All eight began flying in a large circle high above the ridgeline. Shortly thereafter, ravens from the other side of the eastern ridge—from a different territory—began to join our eight. In a matter of minutes, she counted approximately twenty-four ravens flying in the same circle and calling out to one another. The ravens appeared to be greeting and talking to one another, as well as being playful. This event went on for several minutes. Then our ravens flew back into their own canyon, and the other ravens flew out of sight back over the ridge. This was the first time she had seen so many ravens all together at one time. It was as if the adult ravens in this area wanted all of the juveniles to meet and greet one another. Because of a raven's complex intellect, much of what it will need to know for life is not hardwired within its brain at birth. This knowledge must be gained through instruction from the parents. Perhaps making the juveniles aware of other raven families is another of their many lessons.

July 1: This morning was the last time I would be leaving kibbles and chicken scraps before we leave for Denver to spend the Fourth of July with friends. All six of the juveniles were perched in the pine tree and waiting—and squawking. Both adults were in the eucalyptus, taking in the entire scene. Our neighbors volunteered to continue putting food out while we are gone. They have become sufficiently interested in the raven family life over the past few months that they are happy to fill in.

July 5: We returned late this afternoon, just in time for the evening "feeding." At first, all eight ravens were waiting in the pine tree—and squawking. But while I was cutting up some chicken, they left. I called, but no one returned. Our neighbor came by after feeding the other "critters," and while we were talking, the ravens came back.

Although it seemed that the juveniles can feed themselves adequately now, the adults were nonetheless occasionally feeding at least some of them. Even though the majority of the ravens had flown to the pine tree anticipating dinner, the juveniles were hesitant to collect the food. For some reason, the parents didn't seem to mind this time, as both adults hopped down, picked some up, returned to the branches where the juveniles were perched, and fed them.

July 6: Last evening, our neighbor threw some kibbles up onto the balcony. They were still there after dark, but around seven this morning, the juveniles discovered them. All six showed up, and very quickly the kibbles disappeared.

It was a bit foggy today, so the juveniles more or less merely hung around throughout the day, not doing anything special— except maybe squawking. They were taking a well-deserved break from their intensive training. I put out a few extra kibbles for them, and they really enjoyed that. But this evening when I took out the kibbles and chicken pieces, everyone was in the pine tree—all eight! Several of the little ones were still gesturing for the adults to feed them. And the adults accommodated.

July 7: I witnessed a most remarkable phenomenon today in the behavior of the ravens. Sometime in the late morning, I had put out an extra scoop of kibbles for them, and two juveniles and one adult flew down to take advantage of the new supply. But as the two juveniles stepped up to feed on the kibbles, the adult chased one of the juveniles away and would not let it feed until the other had nearly finished. I had never witnessed this behavior before. But as a speculation of what might have happened, perhaps the juvenile that was allowed to feed was one of the youngest who had not yet had an opportunity to fully learn the process, while possibly the other juvenile may have been one of the eldest and may have been acting very greedily in attempting to steal food from its sibling. The adult did ultimately let it feed.

July 8: Our neighbors came over for lunch today, and of course we talked about the ravens. They had recently observed a most curious behavioral anomaly from the ravens in their yard. Through my studies of raven family dynamics, our neighbors have also become interested in the ravens' daily antics and have been putting food on their fence and outdoor tabletop for them. The adults have always stopped by to get it, but now that the juveniles are being taught to fend for themselves, the adults have not allowed them to land on either the fence or the table. Nor have the adults themselves come down to take the food. Both the adults and juveniles know the food is there, but for some reason this source is suddenly "off limits." We think this is another lesson for the juveniles, but we are not sure what the message is.

One possible explanation—which I think is the most plausible—is that the neighbors have a dog. Although the dog has been there all those time the ravens have been picking up the food, it is known to the adult ravens, but not to the juveniles. The adults know that this dog is not a threat. However, the juveniles are now sufficiently mature to be made aware of all forms of danger. The presence of a dog represents potential danger, and this may very well have been another lesson for the youngsters. Until the juveniles themselves became familiar with the dog and its habits, they are not to collect food from this source.

This behavior is most strange, but the raven is such an intelligent bird that this explanation regarding the presence of the dog may very well be the case. I suggested to our neighbors that they put the food on the ground to see if that would make a difference, since the ground is a natural place to find food. (Note: It did not. The ravens never returned to this food resource for the remainder of the juveniles' training.)

July 9: The juveniles are learning very rapidly and very well their critical lessons regarding how to find food—and where to look for it. This morning all six flew over to the pine tree while I cut up some chicken for them. The adults, meanwhile, stayed back to let the little ones feed and fend for themselves. This was all part of their training. After "breakfast," the juveniles went out to practice more aerobatic flights.

The evening was a repeat of the morning. At first, no one was around when I went out to cut up the chicken. I called out to them, and Freya answered and called to the juveniles. While I was chopping the chicken, all six flew over to the eucalyptus tree—two of them squawking incessantly. I put out the kibbles and chicken by the pine tree, and then I called them over. Two of the juveniles flew over right away, and the remaining four followed as soon as I retreated a short distance. Both adults watched for a few minutes and then joined in. The squirrels, who would try to sneak in and steal some kibbles, were greatly outnumbered!

July 10: This morning was a repeat of yesterday. The juveniles waited patiently in the pine tree while I cut up some chicken for them. As I took the kibbles and chicken over to the path at the base of the tree, I called to them. Three flew away, but three stayed in the pine and watched while I put everything down. I stepped back while still calling, and one of them hopped down to investigate. As I walked away, everyone joined in.

In the afternoon I left food out earlier than usual, as we were going to be gone in the early evening. It was a time of day when the ravens weren't normally looking for food, so I didn't call for them. However, when we returned a little after dark, everything was gone.

July 11: I was awakened this morning around six thirty to the sound of squawks and what seemed like hundreds of tiny footsteps pattering on the roof. The juveniles were apparently hungry and looking for food! I didn't get up right then, however. They know the very definition of "early bird." When I finally did get outside with their breakfast, they were all waiting—and hungry!

Occasionally, I see the juveniles on the ground on the hillside across the dry creek, with the adults giving them instructions—and in past years I have observed this same phenomenon with other hatchlings. In this circumstance, the adults may be showing the youngsters how to scratch for food under rocks and rotted vegetation, and/or they may be demonstrating how to hide surplus food. As I mentioned earlier, one of the practices that is characteristic of all ravens is their tendency to store their surplus food by hiding it somewhere in a cache known only to that individual bird. What is not so well known is that if they so much as suspect that a rival raven is watching, they will pretend to store the surplus at the chosen spot, but then secretly remove it to an alternate location.[21] Ravens are so clever at caching that this very well could be a learned behavior.

July 12: The juveniles are really catching on! They will now come over when I call them. And there are two, and often three, who will fly over to the pine tree when I put out kibbles and chicken scraps, and stay there while I'm doing it. It is a thrill watching the whole squadron flying low and slow looking for the goodies! In the meantime, the juveniles have not yet learned the raven language, as they are still communicating with short squawks. The adults have a fairly extensive vocabulary but seem to have no difficulty understanding the juveniles' assorted squawks. One little guy, however, is an extensive squawker and

---

21 T. Bugnvar and K. Kotrschal, "Leading a Conspecific Away from Food in Ravens (*Corvus corax*)," *Animal Cognition* 7, no. 2 (2004): 69–76.

will one day be a real talker. It obviously takes after its mother! It is especially noisy at 6:30 a.m.!

July 13: The juveniles again were up very early this morning, and all of them were outside our window by the pine tree around six thirty calling for breakfast. I did finally get out there about an hour later to take something to them. Everything must be going well in their training, as while I was cutting up some chicken, Freya sat in her favorite perch in the eucalyptus tree and warbled. I called to her and waved back. She continued for several minutes.

I have noticed over the past several days that when I call the ravens at the time I leave food, the juveniles often respond by flying over. They fly low and slow overhead, and usually land in the eucalyptus tree, but occasionally they go directly to the pine where I am standing. This afternoon this behavior was demonstrated quite dramatically. When I put out their kibbles and chicken pieces, no one at all was around. Furthermore, I did not hear any raven calls or juvenile squawks in the background either. I spread out their meal and again called for them. After only my second try, the entire squadron showed up! They all flew in together from the west. A cacophony of ravens was ducking and diving in every direction for several seconds! There was no ambiguity in their minds that it was suppertime. It was an awesome sight!

July 14: The juveniles were up especially early this morning—or it could be that I've finally begun noticing their very early calls. It was 5:30 a.m. when they started making a ruckus outside the window, wondering where breakfast was. That definitely is way too early for me! They had to wait for another couple of hours. They were very impatient today, for when I finally did get out there with their breakfast, no one was around. I called for them and then went to the bird feeders and fish pond to feed those critters. By the time I got to the pond, all the ravens had returned from wherever they had gone.

The juveniles are now flying much farther afield and taking considerably longer flights. They may possibly be nearing the end of their cross-country flight-training period. Lately, when the adults take them out of the area, the entire flock is gone for significantly longer periods. The cadets are very likely getting ready to soon venture out on their own.

July 15: Today we're leaving for a three-week safari trip to Africa to visit the animal herds of Kenya and Tanzania, with a stop at Victoria Falls in Zimbabwe. We will not be returning until the afternoon of August 5. We made these travel plans before I started keeping a daily log of the raven family life. By the time we return, the juvenile ravens will very likely be fairly well finished with their training and ready to leave home. In past years, they typically left sometime around early to mid-August. We do hope to return before everyone has gone, however. Again, our neighbors happily volunteered to feed the ravens in our absence.

August 6: We returned from Africa at 6:00 p.m. yesterday, and I was excited to learn of the status of our raven family. I quickly went around to the rear of the house to see if any were nearby. Initially, none of the ravens was in the immediate vicinity. By the time we got our luggage upstairs, I saw two fly over to the eucalyptus trees. I suspected they were probably the two adults and went outside and called to them. At first there was no response. I took some kibbles over to the pine tree, spread them out, and called again. There was still no response, but then Freya spoke out with her "I am here!" call. When I called back this time, she must have recognized my voice, as she responded with a long series of enthusiastic warbles. Maybe it was my imagination, but it truly did sound like an "I'm glad you're back!" call.

Not realizing we were now home, our neighbor Darlene, who had been feeding the ravens in our absence, came over with some kibbles to give them and saw me talking with Freya. I asked her about the juveniles, as it was obvious from the presence of only the two adults that the juveniles had probably already left home. It is early August and time for them to leave. She confirmed that they had indeed gone. In a way I'd hoped that they might still be around when we returned, but I had suspected that they may not be.

Their training is now complete, and they are out on their own seeking their fortunes. Our neighbor had mentioned in an e-mail we received while in Africa that around July 29 she'd seen only five of the ravens flying together, and they were likely the two adults and three juveniles. Three "graduates" apparently had already left home by that time. But as recently as a day or two before we had returned, all the remaining juveniles had left.

CHAPTER 9

Epilogue to "Family Life in Ravenworld"

● ● ●

So ENDS THE SAGA OF that year's raven family, a brief glance into the window of their daily life. Their world is not all that different from the travails experienced by any other family of any species. The parents are heavily involved with providing for their youngsters' well-being and giving them the best start in life they possibly can. During this intriguing interlude, we witnessed the three stages of development undergone by our little family: egg incubation, hatchlings growing up in the nest, and the training of the youngsters for their ventures out into the world.

During the twenty-five days of egg incubation, beginning with the first day Freya spent the majority of her time on the nest until we identified the arrival of the first hatchling, we observed three significant events. First, although Freya indeed spent virtually her entire day on the nest, she did leave it in the late afternoons for a period of around ten minutes for water and a little exercise. Second, while Freya was on the nest, Tyr regularly brought food to her. And third, which was my greatest surprise, Tyr occasionally took turns with her in performing incubation duties. He would sit on the eggs for upward of one to two hours while she took breaks.

From the time the first hatchling poked its little beak above the nest rim to the moment the first brave fledgling took its "leap of faith" out of the nest, thirty-eight days had passed. During those thirty-eight days, the growth of the little nestlings was phenomenal. We could see almost daily changes in their little bodies as they emerged from helpless, little yellowish blobs with their eyes closed to squirming grayish forms while developing their pin feathers, and on to restless "teenagers" standing at the edge of the nest testing their newly grown flight feathers.

And then their education began. Watching their development for the nearly seventy days of their training was both a delight and lesson all onto itself. Once our little fledglings were out of the nest, their parents wasted no time getting on with the program. Their development progressed step by step in what appeared to be logical increments. It was as if the parents knew what had to be done and how much the little trainees could tolerate and absorb during each stage. Observing their progression from fledglings to competent young adults was for me a most rewarding experience.

The youngsters were now all grown, and their parents had taught them everything they could to get them started in life. And from here on out, how well they would apply their knowledge was now up to them. Life for a young raven is very difficult, and survival is not guaranteed. Studies of marked ravens have shown that as few as half of them would likely survive the first year. Young ravens often tend to fly together in small groups, perhaps for mutual protection, until they find mate and territories. But they will be at least three years old before they actually do mate. For their first few years, they will remain unmated and without territories until one becomes available, typically through the death of a mated raven already in possession of a territory—thus creating an opportunity. Where our six trainees had gone, no one knows. But we do wish them luck!

CHAPTER 10

Tragedy

● ● ●

A TRAGEDY OCCURRED THE FOLLOWING October. The day of the tragedy was an unusual day for October in Southern California. It had been raining here for the majority of the past day and a half, but on that Wednesday it was drizzly with a low-hanging, foggy mist—an eerie foreboding of the sadness to follow.

That morning I had put out a small amount of kibbles and some pieces of chicken scraps as food for the ravens, as I had been doing for the past few years. At that moment both ravens were hunkered down on their favorite limb near the top of the tallest eucalyptus tree east of our house, trying to protect themselves from the dismal weather. I called to them as I had done so many times before and pointed to the food I had left for them. That's all I ever needed to do. These intelligent birds clearly understood. If they were hungry, they would fly from the eucalyptus to the pine tree that stood over where I put the food.

That day, likely because of the weather, they didn't leave their perch right away, so I went back into the house. While most of the time they would fly to the pine tree when called, at times they would not, and I assumed this was one of those times. But once I was in the house, I looked out the patio door and saw that they both had gone over to where I'd put out the chicken bits.

But something about the behavior of one of them wasn't right. While both had ruffled up their feathers to protect against the dampness, one was actively picking up chicken bits, flying away with them, and returning. The other, however, appeared very listless and was moving very little. And it was not eating. I could not tell which raven was ailing. I watched it for a while and then left to do something else.

I came back shortly thereafter to check on it. The raven had moved a few feet to relocate up onto a section of ice plant where Penny had left a few vegetable scraps for the deer and squirrels—and any other creature that would find them of interest. But the raven just sat there, not moving. A deer wandered by and nibbled on the vegetable scraps, but the raven never moved. I watched for a while, and finally the deer walked away. I also left but returned about ten minutes later. When I returned, the raven had moved a few feet farther onto the ice plant and positioned itself facing uphill. The poor bird was definitely ill, and I was helpless to do anything.

As the drizzle increased, the sick raven worked its way over to the shelter of a scrub tree above the ice plant on the hillside and perched itself upon an exposed root. I went out onto the porch and talked to it for a few minutes. It moved its head in recognition, but otherwise remained still. I went back into the house, feeling helpless for not knowing what to do and guilty for not doing anything. It was very sad.

The drizzle was getting heavier as I returned to the house. But I only stayed inside for a few minutes before I checked on the raven again. It was still on its perch under the scrub tree.

But when I checked on it again about fifteen minutes later, it was gone. For a moment I thought that it may have found enough energy to fly away. However, as I walked to the edge of the balcony, I saw that it had fallen off its perch and rolled down the hill. It was lying on its side very still. I immediately went to the garage, donned a pair of rubber gloves, and went outside and picked it up. The poor creature had died.

Penny found a shallow cardboard box from L.L.Bean, and we placed the raven in that. The poor bird appeared to be bleeding slightly from the mouth, as if it had been exposed to rat poison or some other pesticide. We put the body in the garage thinking we could bury it the next day when the weather cleared.

Meanwhile, its mate was flying around the area, making frequent stops and calling for its partner with the characteristic "Where are you?" call. We felt empathy for the mate, as it had no idea of what had just taken place. We had been listening to the mate's call for several minutes when Penny suggested that perhaps we should place the body outside in a location where the mate could see it. We did that, and the next event was the most incredible reaction we have ever seen from a bird.

The box was shallow and roughly a foot square. When I had placed the raven in it, I'd spread its wings out slightly. I took the box over to the location near where we would leave food for the ravens and set it there. The mate was perched in the eucalyptus tree and watched everything. The reaction of the surviving raven when it saw the body of its mate was unbelievable. For over five minutes, it flew over the body low and very erratically with its legs extended, and it called out with a cry we had never heard before. It was a mournfully sad, distressed, almost frantic, hysterical call. It knew its mate was dead. It would perch momentarily to rest and then immediately fly back. It kept returning to check, all the time making that unusual call. The surviving raven was clearly displaying sorrow and grief. These birds are amazingly intelligent!

Very shortly thereafter, two ravens from a nearby territory joined the surviving raven and repeated the same flight behavior with the same call. All three continued this mournful ritual for over a half hour while taking an occasional break in the eucalyptus tree. After an hour or so, we brought the body back into the garage.

We contacted our neighbors and told them the bad news. We planned to bury the raven the following day, as the weather was predicted to be clearing. Our neighbors said that they would like to participate in the memorial.

The following morning the sky was fairly clear. When we returned from our walk, we met up with our neighbors and decided to bury the raven then. Bob had his camera and took several pictures of it before we took it for burial. Bob then mentioned that he'd recently spoken with another neighbor a few houses down who had told him that he had a rat problem in his garage. That neighbor had put out a large quantity of rat poison a couple of days earlier, which he'd told Bob was totally gone. Bob suspected, and I wholly agreed with him, that our raven ate a rat that had died from the rat poison. That would explain the bleeding from the mouth.

Rat poisons are anticoagulant rodenticides, and their use migrates up the food chain, where they kill thousands of predators and birds of prey from internal bleeding. Ironically, the species falling victim to and dying from the effects of these rodenticides are the very predators that keep the rat population in check. This outcome is a very serious unintended consequence. The *Los Angeles Times* reported that recent blood samples taken from 140 bobcats, coyotes, and mountain lions living in the Santa Monica Mountains National Recreation Area revealed that 88 percent of them tested positive for anticoagulant compounds.[22] Fortunately, due to public outcry, public sale of anticoagulants was banned in California in 2014.

We decided to bury the raven at the bottom of our hill above the dry creek. At this point in time, another most remarkable

---

22 Louis Sahagun, "War on Rats Claims Other Casualties," *Los Angeles Times*, February 8, 2016.

phenomenon occurred. As we picked up the raven in the box and started over the hill, approximately eight to ten ravens gathered overhead, circled over us for about five minutes, and then flew on. Was this a coincidence? Did they know of the death of one of their comrades? How did they know? There have been numerous anecdotal accounts of this behavior in ravens, but this was my first experience actually witnessing it.

We buried the raven and marked its grave. We decided that any species this intelligent deserves a marking. At the time of the tragedy, we all had been friends for over eighteen years. Very shortly thereafter—that same day, as I remember—the surviving raven left the area, and we no longer saw it anywhere in or around the canyon. And we had no idea which raven, Freya or Tyr, we had buried.

CHAPTER 11

A New Beginning

● ● ●

CONSIDERABLE TIME HAD PASSED, PERHAPS a month or two, since we'd lost our ravens, and Snover Canyon was eerily silent from their missing calls. I missed the daily interactions I'd had with these remarkable beings. I felt so fortunate to have been given the rare opportunity to experience their presence and share in their lives.

One morning during this period of their absence, I set out for my daily walk. The air was cool and fresh, and the skies were clear. As I was heading west down our fairly long driveway and past our neighbors' house, I approached the first row of pines that lined the street. As I neared the first pine, a raven flew to it and landed on one of the lower branches. It then turned to face me and warbled! I recognized that warble. It was Freya! She had returned! Until that moment, I hadn't known which of the two ravens had died.

I waved to her and called back. She warbled again. We communicated for at least another minute before she left the branch and flew over our house toward the eucalyptus trees. I quickly followed and walked around to the back of the house. She had landed at her favorite perch near the top of the largest one. As I came around the corner, she warbled again. I truly believe she was happy to see me—as happy as I was to see her! I waved and talked to her.

As I was doing that, another raven approached the eucalyptus copse from the far side and landed on one of the adjacent branches. Freya had found a new mate and had brought him to her territory. What a wonderful surprise! The canyon will once more echo with their calls. Again, we will delight in their presence.

I will once more be leaving out chicken bits and kibbles, and I am sure Freya will show her new beau what that process is all about. He will be very shy and standoffish for a while. But as he learns from Freya that I pose no threat, he will soon become less intimidated. We all are about to embark on a new adventure. I can hardly wait to see where it will take us!